W0113295

INDIA AND THE GLOBAL GAME
OF GAS PIPELINES

Gas pipelines constitute an important, yet unexplored, aspect of strategic geography. As one of the fastest-growing economies in the world, India's need for energy is paramount. Though surrounded by gas-rich regions – Myanmar and Bangladesh to the east, the Gulf to the west and Central Asia to the north – India does not have a single gas pipeline coming in, going out or traversing through its territory to date.

This book highlights the global competition over gas pipelines and its implications for India's energy security in a comprehensive manner. The author leads us through a labyrinthine world comprising numerous actors – the states, energy firms, scientists, engineers, investors and bankers – engaged in competition over these pipelines leading to a continuous game of checkmating rivals, instigating conflicts, causing damage and destruction and threatening military action to persuade or dissuade states from joining specific projects.

Pulsating, rigorous, grounded in hard facts and solid research, this book will be indispensable for scholars and researchers of international relations, strategic affairs, defence studies and politics, as well as think tanks, government agencies and the informed general reader.

Gulshan Dietl is former Professor, School of International Studies, Jawaharlal Nehru University, New Delhi, where she also served as the Director of the Gulf Studies Programme and the Chairperson of the Centre for West Asian and African Studies. She was Fulbright Scholar-in-Residence at the Mount Saint Mary College, Newburgh, New York (1993–1994), Guest Research Fellow at the Copenhagen Peace Research Institute (1998–1999), Visiting Professor at the University of Kashmir (2004), Associate Director of Research at the Fondation de la Maison des Sciences de l'Homme, Paris (2008), Visiting Professor at the University of Southern Denmark (2010), Visiting Professor at Jamia Millia Islamia, New Delhi (2012–2013) and ICSSR Senior Fellow at the Institute of Defence Studies and Analyses (2013–2015).

'The great gas game is afoot, and its consequences are vital for India's energy security. . . . Gulshan Dietl covers largely unchartered territory by analysing the pipeline politics of gas from the Persian Gulf to South Asia.'

Luke Pate, Danish Institute for International Studies and
Oxford Institute for Energy Studies

'Dietl's foray into gas pipelines and their geopolitics is a timely effort that will benefit scholars grappling with a complex subject like energy security. . . . The book fills a vacuum in the literature on energy security.'

Sudha Mahalingam, independent energy consultant and
former energy regulator

INDIA AND THE GLOBAL GAME OF GAS PIPELINES

Gulshan Dietl

Routledge
Taylor & Francis Group

LONDON AND NEW YORK

First published 2017 by Routledge

2 Park Square, Milton Park, Abingdon, Oxfordshire OX14 4RN
52 Vanderbilt Avenue, New York, NY 10017

*Routledge is an imprint of the Taylor & Francis Group, an
informa business*

First issued in paperback 2019

Copyright © 2017 Gulshan Dietl

The right of Gulshan Dietl to be identified as author of this
work has been asserted by her in accordance with sections 77
and 78 of the Copyright, Designs and Patents Act 1988.

The international boundaries, coastlines, denominations,
and other information shown in any map in this work do not
necessarily imply any judgement concerning the legal status
of any territory or the endorsement or acceptance of such
information. For current boundaries, readers may refer to the
Survey of India maps.

All rights reserved. No part of this book may be reprinted
or reproduced or utilised in any form or by any electronic,
mechanical, or other means, now known or hereafter invented,
including photocopying and recording, or in any information
storage or retrieval system, without permission in writing from
the publishers.

Notice:
Product or corporate names may be trademarks or registered
trademarks, and are used only for identification and explanation
without intent to infringe.

British Library Cataloguing in Publication Data
A catalogue record for this book is available from the British Library

Library of Congress Cataloging-in-Publication Data
A catalog record has been requested for this book

ISBN: 978-1-138-23546-5 (hbk)
ISBN: 978-0-367-27608-9 (pbk)

Typeset in Galliard
by Apex CoVantage, LLC

CONTENTS

CONTENTS

FIGURES

ABBREVIATIONS

APEC	Asia-Pacific Economic Cooperation
BCF	billion cubic feet
BCM	billion cubic metres
BNP	Bangladesh Nationalist Party
BRICS	Brazil, Russia, India, China, South Africa
BTC	Baku–Tbilisi–Ceyhan
BTE	Baku–Tbilisi–Erzurum
CCGT	combined-cycle gas turbine
CIS	Commonwealth of Independent States
CNG	compressed natural gas
CNOOC	China National Offshore Oil Corporation
CNPC	China National Petroleum Corporation
DRAs	drag reduction agents
EDF	Électricité de France
EEZ	exclusive economic zone
EGL	Elektrizitaets-Gesellschaft Laufenburg
EIA	Energy Information Administration
ENI	Ente Nazionale Idrocarburi (national hydrocarbons authority). The acronym is now the official name.
E.on	From the Greek aeon, meaning 'eternity'
FPSO	floating production, storage and offloading
FSU	Former Soviet Union
GAIL	Gas Authority of India Limited
GAZPROM	Gazovaya Promyshlennost
GBC	Gas Buyers Club
GCA	Gaffney, Cline and Associates
GECF	Gas Exporting Countries Forum
GSPA	gas sales purchase agreement
GSPC	Gujarat State Petroleum Corporation

IBRD	International Bank for Reconstruction and Development
IEA	International Energy Agency
ILSA	Iran, Libya Sanctions Act
IOC	International Oil Corporations
IRNA	Islamic Republic News Agency
ITGI	Interconnector Turkey, Greece, Italy
JPC	joint project company
LNG	liquefied natural gas
LPG	liquefied petroleum gas
MBI	Myanmar–Bangladesh–India
MCF	million cubic feet
MCM	million cubic metres
MoPNG	Ministry of Petroleum and Natural Gas
NATO	North Atlantic Treaty Organisation
NELP	New Exploration and Licensing Policy
NGO	non-governmental organisation
NIOC	National Iranian Oil Company
OAPEC	Organisation of Arab Petroleum Exporting Countries
OIL	Oil India Limited
OMV	Österreichische Mineralölverwaltung. The acronym is now the official name.
ONGC	Oil and Natural Gas Corporation
OPEC	Organisation of Petroleum Exporting Countries
OSCE	Organisation for Security and Cooperation in Europe
PEACE	Pipeline Extending from the Asian Countries to Europe
PKK	Partiya Karkeren Kurdistan (Kurdistan Workers Party)
PNGRB	Petroleum and Natural Gas Regulatory Board
RBC	Royal Bank of Canada
RGIHV	Report of the Group on India Hydrocarbon Vision
RGTIL	Reliance Gas Transportation Infrastructure Limited
RIL	Reliance Industries Limited
RWE	Rheinisch-Westfälisches Elektrizitätswerk. The acronym is now the official name.
SAARC	South Asian Association for Regional Cooperation
SAIG	South Asia Integrated Gas
SAIPEM	Società Anonima Italiana Perforazioni E. Montaggi. A subsidiary of ENI, its acronym is now the official name.
SCO	Shanghai Cooperation Organisation
Sinopec	China Petroleum and Chemical Corporation
TAGP	Trans-Afghan Gas Pipeline
TANAP	Trans-Anatolian Gas Pipeline

TAP	Trans-Adriatic Pipeline
TCP	Trans-Caspian Pipeline
toe	tonne of oil equivalent
TRACECA	Transport Corridor Europe–Caucasus–Asia
UNCLOS	United Nations Convention on the Law of the Sea
WTO	World Trade Organisation

INTRODUCTION

> Virtually all current geopolitical developments are energy related.[1]
>
> Behind the shift in strategic geography is the new emphasis on the protection of supplies of vital resources, specially oil and natural gas. Whereas Cold War era divisions were created and alliances formed along ideological lines, economic competition now drives international relations and competition over access to these vital economic assets has intensified accordingly.[2]

The quotes here put energy at the very heart of geopolitics and draw attention to the shift from ideology to economic competition for the energy resources. The chapter proposes to introduce the theme of the research project.

Statement of the problem

The territorial state has come to stay within its sovereign space inside the well-defined borders visible on the map. Alongside these borders, there are invisible additional lines signifying trading blocks, military alliances, regional organisations, multinational corporations, international NGOs and a plethora of cross-border digital communities. Together with these supra-state entities, there are ethnic, sectarian and linguistic sub-state identities that seek to etch themselves on the state territory. The energy pipelines, carrying oil and gas from the well head to the market, generally run through two or more states, and often from one continent to the other. Even as they constitute an important aspect of strategic geography today, the pipelines have remained the least examined of the spatial entities in geopolitical terms.

The geopolitics is all about pursuing a favourable strategic geography that would facilitate control over or access to the resources of the planet. Water, agriculture and animals on the ground; marine resources in the water; and mineral and metal resources under the ground and under water ensure life support. The roads, railways, ships, aeroplanes, bridges, beaches, rivers, lakes, harbours, straits and islands lead to these resources. The energy pipelines are the coveted objects of geopolitics for three distinct reasons: for the commodity they contain, as the containers of that commodity and as the carriers of the commodity. Each of the three – the resource, storage and passage – is vital in itself. An energy pipeline demands tough decisions and difficult implementation.

The gas pipelines deserve a closer look for several reasons. One, more than 90 per cent of gas produced and three-quarters of gas traded is distributed by the pipelines. Two, while the Organisation of Petroleum Exporting Countries (OPEC) sets the price and production of oil, gas has no equivalent body. It is traded bilaterally and multilaterally. A gas pipeline, therefore, is much more prone to political influences than an oil pipeline. The route, direction, volume, length, thickness and width are not only technological and commercial decisions; they are political decisions taken at the top of the political hierarchy in a state. Long-term investment and huge infrastructure are prerequisites to an agreement to construct a gas pipeline. Three, the share of gas is going up in the global energy basket. Four, gas has a much longer production profile than oil. And five, as the production sites move to less accessible and landlocked areas, pipelines become the only feasible transport to move them to the market.

Theoretical and conceptual contexts

There is a plethora of literature on the links between resources and wars, both intrastate and interstate. The scarce, precious and depleting resources are particularly prone to becoming the sites of contestation and conflicts. The states, energy firms, scientists, engineers, investors and bankers make up the entire gamut of the gas world. There is fierce competition between the states engaged in the gas trade, as also among the gas entities within the states. Whereas cooperation is conditioned by common interests, competition leads to instigating conflicts, causing damages and destructions, checkmating the rival projects and, at times, also threatening military action to persuade or dissuade a state from joining a project. An open war fought by two adversary armies has not happened – so far. Since it is neither small in volume nor easy to extract, the non-state actors are not involved in stealing or smuggling, and their role remains confined to sabotage.

The larger context within which the study is undertaken is the states' search for energy security. Energy plays an important role in the national security of any given country, although the energy quotient in the national security calculations may differ from one country to the next. The International Energy Agency (IEA) defines energy security as the uninterrupted availability of energy sources at an affordable price. Energy security has many aspects: long-term energy security mainly deals with timely investments to supply energy in line with economic development and environmental needs. On the other hand, short-term energy security focuses on the ability of the energy system to react promptly to sudden changes in the supply–demand balance.[3] A wider definition of energy security goes beyond sustainability, competitiveness and secure supply. It is a multidimensional concept, including external as well as internal actions. Economic, political and security measures have to be applied in combination to generate the essential synergies. Thus only an integrated approach that combines all the different aspects of energy security can be successful.[4] Human resource development, development of energy infrastructure, constant technological upgrades, diversification of supplies and markets, investments and fuel substitutes are indispensable ingredients of energy security.

To guard against the sources of insecurity is as important as to take positive measures listed earlier. The threat to energy security could come from the competing states, more often though, from natural and human actions. The damage to the energy infrastructure from human intervention could be either accidental or motivated. The terrorist attacks, in turn, could be the result of individual or group action borne out of a variety of grievances targeting the beneficiary of the energy project. The sources of insecurity are multiple and multiplying with time.

Realism dominated the Cold War years, as it provided explanations for wars, alliances, containment and a constant deterrence between the two superpowers. The theory has continued to evolve since then. Staying within the mainstream neo-realism, defensive realism would be better applicable to understand the driving impulses of the actors in the gas game. Kenneth Waltz, *Theory of International Relations* (Addison-Wesley, Reading, 1979) continues to remain the foundational text of defensive realism long after its publication. Unlike the offensive realists, who argue that the states should always be looking for opportunities to maximise power and aim for hegemony, the defensive realists maintain that the states should strive for appropriate power. Faced with hegemony, the states should build up their own capabilities (internal balancing) and form a balancing coalition (external balancing). Absolute power leads to wars.

Can defensive realism be stretched to incorporate cooperative behaviour, when the outcome is sum–sum? Is it also possible to acknowledge non-state actors, although realism privileges the self-centred, self-contained and sovereign state as the central actor in the international relations? Is it possible to take on board domestic militarism and hyper-nationalism as influencing factors in state behaviour in addition to international anarchy?

As the quotes at the beginning suggest, the theme involves a global commodity and its pursuit by state and non-state actors alike. The study, therefore, will need to borrow from more than one theoretical constructs. Realism is one of several possible frameworks, better applicable in certain times and situations than in others. The globalists assert, on the other hand, that the state is receding. It is no longer the exclusive actor on the global scene. For example, Susan Strange questions the role of the state in *The Retreat of the State: The Diffusion of Power in the World Economy* (Cambridge University Press, Cambridge, 1996). Mainly, globalism is an economic, rather than political, phenomenon. It deals with cross-border flows of goods, investments, information, tourists, refugees, migrants, environment, diseases and so on. A commodity like natural gas flowing through international and intercontinental pipelines must necessarily take a broad angle view of the world. The global gas resource, the production profile, the expanding/shrinking/shifting market and the price fluctuations can only be looked at from a global viewpoint.

Looked at from the realist point of view, would the state go to war to ensure energy security? So far, it has stopped just short of active military engagement. Is that to be explained in terms of the dichotomy between vital existential issues and the non-strategic ones? Although the energy pursuit is seen as a resource war, it remains within the field of 'low politics' even as fierce contestation goes on among the stakeholders. The 'high politics' was the central focus of security studies during the Cold War and dealt mainly with issues of state survival and sovereignty. Since the Cold War, security is reconfigured to encompass military and economic aspects. Some analysts hold that the low politics has now graduated to high politics. Others consider the distinction a false dichotomy as there is a complex interdependence between the two realms. Robert O. Keohane and Joseph S. Nye in *Power and Interdependence: World Politics in Transition* (Little Brown and Company, Boston, 1977) hold that the international relations have evolved from a simple scheme based on national security to a complex interdependence based on a number of issues. For the purpose of the study, access to energy, trade, environmental pollution and international drug trafficking are considered four major non-military components of security. Together, they call for a coordinated approach comprising economics, diplomacy and defence.

Geography, strategic geography and geopolitics are related terms and relevant to the study. The globalists' claim that geography is history has proved to be too premature. A central variable in the life of a state, geography deals with physical features of the planet. Napoleon is said to have remarked: 'A country's geography is its fate.' The truculent personality of Prussia and Czarist Russia had much to do with their being land powers with few natural borders to defend them, whereas the United Kingdom (UK), the United States (US) and Venice, Italy, could champion liberty because they had the luxury to be protected from meddlesome neighbours by expanses of surrounding water.[5] One does not have to accept geographical determinism, but one cannot reject geography as an important ingredient in international relations. In fact, Robert D. Kaplan makes a strong plea to seat the geographical determinists at the same honoured table as liberal humanists, as geography in the most old-fashioned sense is making a comeback in his seminal article 'The Revenge of Geography', *Foreign Policy*, May–June 2009.

Strategic geography focuses on the control of or access to the spatial areas that are indispensable to the survival, security and prosperity of state. Zbigniew Brzezinski, in his seminal book *The Grand Chessboard: American Primacy and Its Geostrategic Imperatives* (Basic Books, New York, 1997), prescribed three grand imperatives of imperial geostrategy: prevent collusion and maintain security dependence among the vassals, keep tributaries pliant and protected and keep the barbarians from coming together. The worst-case scenario, according to him, is a grand coalition of China, Russia and perhaps Iran, an 'anti-hegemonic' coalition united not by ideology but by complementary grievances.

Geopolitics, since its very coinage, has been a fashionable and an esoteric concept. And yet, it remains relevant to the study. Traditionally, it meant causal relationship between the political power and geographical space. Today, war and geopolitics are likened to Siamese twins or at least to inseparable lovers by Filip Tunjic, 'War and Geopolitics – Really together Again?', *Strategic Digest*, January 2000, pp. 20–34. Klaus Dodds, *Global Geopolitics: A Critical Introduction* (Pearson Education Ltd, Essex, 2005) concludes his study by pointing to the Pentagon conceptualisation of three areas – Caspian Sea basin, South China Sea and West Asia – as part of a gigantic 'strategic triangle', which will shape the pattern of potential wars in the twenty-first century. Two of the three identified possible war zones are contiguous, are rich in gas and constitute the focal point of the study. Charles Clover, 'Dreams of the Eurasian Heartland: The Re-emergence of Geopolitics', *Foreign Affairs*, March–April 1999. https://www.foreignaf fairs.com/articles/asia/1999–03–01/dreams-eurasian-heartland-reemer gence-geopolitics derides war-inducing aspect of the concept, saying, 'Few modern ideologies are as whimsically all-encompassing, as romantically obscure, as intellectually sloppy, and as likely to start a third world war as the theory of geopolitics'.

Cartel is yet another relevant concept. Neither the gas sellers nor the gas buyers have been able to come together to form an exporters' or a buyers' cartel so far. Neither, however, has called it quits. In economic terms, cartel is an agreement between competing firms to control prices or exclude the entry of a new entrant. Arthur Sullivan and Steven M. Sheffrin, *Economics: Principles in Action* (Pearson Prentice Hall, New Jersey, 2003) define it as a formal organisation of sellers or buyers that agree to fix selling prices, purchase prices or reduce production using a variety of tactics. Internal cartels of firms are illegal and prosecuted under antitrust laws in most countries. OPEC, Organisation of Arab Petroleum Exporting Countries (OAPEC) and International Energy Agency (IEA) are state cartels and thus beyond the authority of state laws. The gas buyers' and gas sellers' cartels, if constituted, would fall under this category.

It is economic terrorism which is bloodless but as effective. Since the 9/11, there is an increasing focus on inflicting economic costs on the enemy by damaging and destroying economic assets. Terrorism is as much about causing harm as about demonstrating its presence and purpose. An explosion in the pipeline results in losses, and it creates a loud sound accompanied by fire. Brian A. Jackson, Lloyd Dixon and Victoria A. Greenfield, *Economically Targeted Terrorism: A Review of the Literature and a Framework for Considering Defensive Approaches* (Rand Corporation, Santa Monica, 2007) concludes that an act of economic terrorism cannot bring a major economy to its knees. It does, however, mean immense precautionary measures and substantial expenses to prevent its occurrence. And restoring a pipeline after the attack results in more expenses, even while the consumers are left stranded. Often, the project is given up, half way through its construction, in the face of such a threat.

Plan of study

The first two chapters are devoted to a detailed study of the gas and the pipelines. Chapter 1 seeks to explain the composition, history, geographical spread/concentration, its place in the energy basket and the trends at regionalising/globalising the gas trade. It looks at the future prospects of gas in view of the shale revolution and the recent discovery of gas in the eastern Mediterranean.

Chapter 2 takes up the gas pipelines in general for a closer look. The value of a pipeline is three-fold: for the commodity, as a container and as its conveyer. It is the end result of sheer physical work, immense financial commitment and the most complex technological input. Additionally, it may involve hard bargains, disputes and conflicts.. Today, 95 per cent of produced gas and 75 per cent of traded gas is transported via transnational pipelines. The rest is transported as the liquefied natural gas (LNG) carried in the cryogenic containers. Natural disasters, accidental human interventions and terrorist attacks leave the pipelines vulnerable.

The introductory chapters are followed by three case studies of the major pipeline gas–exporting countries, that is Iran, Russia and Turkmenistan. Qatar is an LNG gas exporter and, therefore, does not come within the scope of the study. Chapter 3 is a study of Iran's gas resources and its policies on gas production and exports. It is to be noted that contrary to the general perceptions, it is Iran and not Russia that has the maximum stocks of gas. It also holds the second-largest stock of oil after Saudi Arabia. Its location on the energy-rich Gulf, the Caspian and the Strait of Hormuz gives it a unique place in the global energy trade. Most of its pipeline initiatives have so far been frustrated. Today, the sanctions on Iran are already unravelling. The Iranians know this and are all set to open up the gas spigot in full force. There is hope that the Nabucco pipeline – long in hibernation – may be resurrected with gas inputs from Iran.

Chapter 4 is devoted to Russia, again a country under sanctions. With the second-largest stock of natural gas in the world, Russia plays a pivotal role not just because it holds vast gas resources but also because it sits astride the large European and emerging Chinese markets. Gazprom, the Russian energy 'behemoth', is the largest extractor of natural gas in the world that controls more than 60 per cent of gas reserves and 80 per cent of gas production in Russia. Two major pipelines from the country are examined in detail. The South Stream pipeline that was planned to circumvent Ukraine had to be dropped as Bulgaria pulled out of the project. It has now been resurrected as the Turkish pipeline will run through it. The two mega pipeline projects with China have finally been signed after a decade-long hiatus.

Chapter 5 deals with the Turkmenistan – a reclusive state with a small population and a short history. Its landlocked location and the poor quality of its gas notwithstanding, the country has successfully managed to enhance its reserves and attempts to diversify its exports. The Turkmen gas policy rests on two pillars: one, it sells its gas at the border and takes no responsibility for the safe passage beyond it, and, two, it permits only the domestic companies to develop onshore fields. An exception was made in the case of the China National Petroleum Corporation (CNPC) and much later in the case of the Italian company Eni. The Turkmenistan–Russia–Ukraine pipeline is a constant source of friction and one of the causes of the ongoing civil war/de facto division of Ukraine. It supplies gas to China through the Central Asia–China line, which is the longest pipeline in the world.

The organisation of the case studies has followed a simple principle. Since most of these lines pass through more than one of these states, each one is studied at the point of its origin. Thus, the Turkmenistan–Russia–Ukraine line is studied in the chapter on Turkmenistan even though Russia has a much higher stake in the line and is a much higher-profile actor. The chapter on Russia looks at the line as an example of the country leveraging its energy to achieve its broad strategic goals. The pipeline itself is studied in the chapter on Turkmenistan.

As per the self-set guidelines, the Iran–Pakistan–India pipeline should have been studied under the Iran chapter, where it was to originate. Instead, it has been taken up in the India chapter, where it was scheduled to terminate. Similarly, Turkmenistan–Afghanistan–Pakistan–India pipeline should have been studied under the Turkmenistan chapter, where it would originate. It has also been taken up in the India chapter. An explanation is necessary. It is because the entire project is geared towards the situation of gas woes and its possible solution in India.

There is yet another obvious exception – India is the only country that is not a gas exporter and has yet been included in the study.

Sources and methodology

I have been engaged with energy studies as a student of West Asian studies over long decades. This is the first full-length study on the energy that I have been able to undertake. The Indian Council for Social Science Research (ICSSR) granted me a senior fellowship to pursue the study and the Institute for Defence Studies and Analyses (IDSA) granted me affiliation and extended all possible facilities to make my stay there pleasant and fruitful. I have benefitted immensely from the IDSA library stacks and the e-resources.

The Indian Ministry of Petroleum and Natural Gas hosts a regularly updated website at http://petroleum.nic.in. The Oil and Natural Gas Commission has an equally useful website at http://www.ongcvidesh.com. Both together have provided statistics, chronology and facts that I have accessed and utilised. The bibliography at the end of the study is a testimony to the ocean of database that one can delve into.

I was fortunate to have visited Russia, Iran and France prior to joining the fellowship. The visits were in connection with the independent research on the same theme. The scholars and policymakers I met and the written literature that I collected were invaluable inputs in the study. In Moscow, I would particularly like to mention Professor Vladimir Feigin, the president of the Institute for Energy and Finance; Professor Nodari Simonia, Deputy Director of the Institute of World Economy and International Relations (IMEMO); and Dr Tatiana Mitrova, head of the Oil and Gas Department at the Energy Research Institute, who were forthcoming with facts and analyses. In Tehran, I would like to acknowledge Dr Narsi Ghorban, the managing director of Narkangan Gas to Liquid International Company and academic author, for sharing his insights and writings with me. Dr Krzysztof Kwiecien, the natural gas analyst at the International Energy Agency in Paris, provided an insightful overview of the gas market. Durdyyev Parahat Hommodovich, the Turkmen ambassador in India, was convincing in his argument for the Turkmenistan–Afghanistan–Pakistan–India pipeline. The list of scholars, energy experts, friends and well-wishers in India is long. I desist from naming them out of fear that I may miss out some of them. They know their contribution to the study and understand my silent appreciation.

Notes

1 Pepe Escobar, 'Playing Chess in Eurasia', *Opednews*, 25 December 2012. http://opednews/articles/Playing-Chess-in-Eurasia-by-Pepe-Escobar-111225-365.html.
2 Michael T. Klare, *Resource Wars: The New Landscape of Global Conflict* (Henry Holt and Company, New York, 2001).
3 'The IEA's Role in Global Energy Security', *International Energy Agency*. http://www.iea.org/topics/energysecurity. Accessed on 15 June 2014.
4 Florian Baumann, 'Energy Security as Multidimensional Concept', *Center for Applied Policy Research Policy Analysis*, no. 1, March 2008. http://edoc.vifapol.de/opus/volltexte/2009/784/pdf/CAP_Policy_Analysis_2008_01.pdf. Accessed on 13 January 2014.
5 Robert D. Kaplan, 'Actually, It's Mountains', *Foreign Policy*, July–August 2010, p. 105. Also Ibn Khaldun, Montesquieu, Adam Smith and Andre Siegfried viewed spatial relationships as the sine qua non of any understanding of the social, economic and political development of human societies. Indeed, Smith identified geography along with the free market as the two key determinants of a nation's wealth. Recent studies employing geography as an explanation for patterns of development have overcome past limitations and criticisms by treating it as a primary explanatory factor but not as an over determined one. Dwayne Woods, 'Bringing Geography Back in: Civilisations, Wealth and Poverty', *International Studies Review* (Malden), Vol. 5, no. 3, September 2003, p. 344.

Part I

RESOURCE AND ROUTES

1

NATURAL GAS
Geology, geography and markets

Unlike coal mining, the production of natural gas does not beget strip-mined wastelands nor does it foster black-lung disease. And unlike petroleum, it does not sully beaches and birds with the products of its transport mishaps. Natural gas does not threaten the wilderness and recreational values of free flowing streams nor does it impose the multi-generational responsibilities of nuclear waste. Combustion of gas is relatively free of soot, carbon monoxide, and the nitrogen oxides associated with other fossil fuels.[1]

Even though we have revised our growth estimates downwards, the 'Golden Age' of gas remains in full swing. Gas is already a major fuel in power generation, but the next five years will also see it emerging as a significant transportation fuel, driven by abundant supplies as well as concerns about oil dependency and air pollution. Once the infrastructure barriers are tackled, natural gas has significant potential for clean-energy use in heavy-duty transport where electrification is not possible.[2]

The quotes rank the natural gas as the best among the coal, oil, gas and nuclear fuels and forecast its continuing and expanding utility in coming years as the best energy source. The chapter proposes to look at this commodity in detail in all aspects.

Composition and early discoveries

Natural gas consists mainly of methane, which is one atom of carbon and four atoms of hydrogen (CH_4), the simplest hydrocarbon, along with more complex natural gases such as ethane (C_2H_6), propane (C_3H_8) and butane (C_4H_{10}). It is the cleanest-burning fossil fuel, producing mostly just water

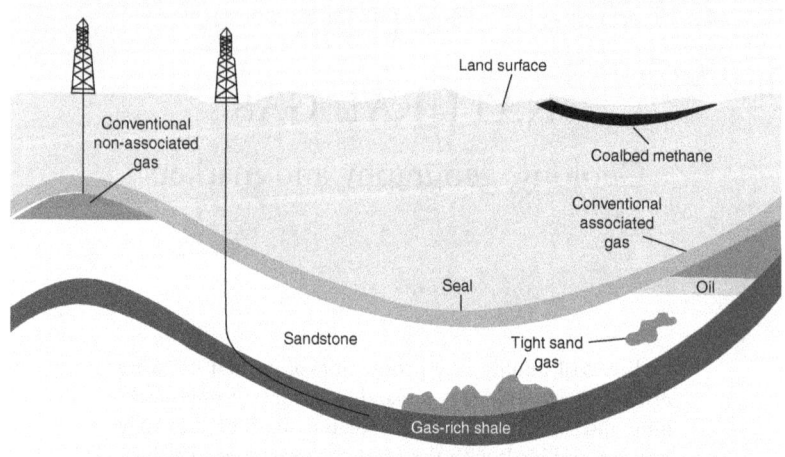

Figure 1.1 Schematic geology of natural gas resources.

Source: Based on US Energy Information Administration.

vapour and carbon dioxide.[3] Methane is also a key raw material for making solvents and other organic chemicals. The proportion of methane in natural gas is typically 85–95 per cent, depending on the location of the well head.[4] Methane can be converted to methanol, an alcohol, for use as a gasoline substitute. Propane and butane are usually extracted from natural gas and sold separately. LPG, which is mainly propane, is a common substitute for natural gas.

There are several schools of thought on the origin of natural gas on the planet. Most, but not all, petroleum geologists adhere to the 'classical' hypothesis regarding the origin of gas. For oil or gas to occur in the subsurface, a certain number of processes must have happened in antiquity. First, there must have been a collection of organic material, the carbon source, such as plant or animal material, concentrated in such a way that it eventually became a layer of rock. Temperatures and pressures in this rock must have been high enough to have 'cooked' the organic material into precursors of fluid hydrocarbons. The hydrocarbons might have remained in the source rock but usually would have moved from the high-pressure environment in which they formed to lower-pressure regions closer to the surface. Then this hydrocarbon migration through rock pores and cracks must have stopped against some impermeable rock layer where the hydrocarbons remain today. An alternative conjecture about the sources of crude oil and natural gas attributes them to the constant upward percolation of inorganic methane from the earth's core under enormous pressures. After migrating past a nearly impermeable zone of highly compressed rock several kilometres deep, the theory states, part of this methane comes to rest where it can be found by explorationists.[5]

The either-or origin of the gas is not universally accepted. An inclusive understanding of the phenomenon of gas traces its origin to several processes.[6] The above-mentioned first hypothesis refers to the fossil gas that was formed a few hundred million years ago from the decomposed remains of plants and primitive life forms that got deposited at the bottom of ancient lakes and oceans. The above-mentioned alternative conjecture refers to the thermogenic gas that is formed at lower depths and at higher temperatures (above 150 degree Celsius) beneath impermeable rock formations. Apart from these two, a third source of gas is minerals that originated billions of years ago, when the planets coalesced from stardust. It is found throughout the solar system in this form. Fourth, the biogenic or bacterial gas is constantly being created by the bacteria inside the earth. At relatively shallow depths where the temperatures are not high enough to generate oil, bacterial action quickly produces biogenic or microbial gas. Commonly known as swamp or marsh gas, it is rarely contained; instead, it leaks out into the atmosphere in enormous volumes. This gas is almost pure methane. Urengoy in Siberia,[7] which is one of the largest gas fields in the world, is biogenic in origin.

The gas discovery begins with the geologist, who locates the gas-rich terrain, nowadays with the help of seismic devices and satellite imagery that indicate the presence of gas below surface. The emerging gas industry in the US and Europe was based not on natural gas but on 'manufactured' gas, which was made by heating coal.[8] It was in Pazanan in Iran where gas was first discovered in 1937, followed by small discovery in St Marcet in France two years later. The first major discovery was the Harsi R'Mel field in Algeria that contained the reserves of 100 trillion cubic feet (tcf, i.e. 2.8 trillion cubic metres – tcm) and Surt basin in Libya in 1956. This was closely followed by similar finds in Abu Dhabi, Saudi Arabia, Nigeria, Lacq field in France and Slochteren field in Groningen Province in Holland. The mid-1960s witnessed discoveries of the West Sole in the North Sea Southern basin, followed rapidly by Leman, Indefatigable, Hewitt, Rough and Hamilton in the United Kingdom. Later in the decade, the giant potential of Russian gas reserves began to be realised and Alaska emerged on the gas map as also Mexico. Thus, the great period of gas discovery began in the mid-1950s, peaked in 1960s and extended throughout in 1970s.[9]

Table 1.1 Volume of gas reserves

At the end of 2013, the country-wise proved reserves of natural gas in tcm stood at:

Natural Gas Proved Reserves at the End of 2013 in Trillion Cubic Metres[10]	
US	9.3
Canada	2.0
Mexico	0.3
Argentina	0.3
Bolivia	0.3
Brazil	0.5
Columbia	0.2
Peru	0.4
Trinidad and Tobago	0.4
Venezuela	5.6
Other South and Central America	0.1
Azerbaijan	0.9
Denmark	. . .
Germany	. . .
Italy	0.1
Kazakhstan	1.5
The Netherlands	0.9
Norway	2.0
Poland	0.1
Romania	0.1
Russia	31.3
Turkmenistan	17.5
Ukraine	0.6
United Kingdom	0.2
Uzbekistan	1.1
Other Europe and Eurasia	0.2
Bahrain	0.2
Iran	33.8
Iraq	3.6
Kuwait	1.8
Oman	0.9
Qatar	24.7
Saudi Arabia	8.2
Syria	0.3
United Arab Emirates	6.1
Yemen	0.5
Other West Asia	0.2
Algeria	4.5
Egypt	1.8
Libya	1.5
Nigeria	5.1
Other Africa	1.2
Australia	3.7

Natural Gas Proved Reserves at the End of 2013 in Trillion Cubic Metres[10]

Bangladesh	0.3
Brunei	0.3
China	3.3
India	1.4
Indonesia	2.9
Malaysia	1.1
Myanmar	0.3
Pakistan	0.6
Papua New Guinea	0.2
Thailand	0.3
Vietnam	0.6
Other Asia-Pacific	0.3
Total World	185.7

The estimates in Table 1.1 are from the British Petroleum, which is a trusted source of energy statistics. There are several other widely-referred-to sources doing similar exercises. Their methodologies may differ, the units they delimit for scrutiny may be different and the sources they rely upon may not be the same either.[11] The numbers do not tally in all sources, but neither are they far apart. And these estimates are just that – estimates. They can be proven to be wrong in either underestimation or overestimation. The faulty estimates could be the result of genuine mistake or of motivated arithmetic. Today's proved reserves are several times larger than a few years ago, in fact, two to six times higher. As new discoveries come on ground, as new technologies become available and as better methods of transport are devised, larger volumes of gas will become exploitable. On the other hand, with other variables remaining constant, the gas consumption will deplete the reserves and drive the estimates downwards.

Some gas reservoirs do not contain any liquid hydrocarbons. The gas produced from these reservoirs is called dry gas or non-associated gas. Many other reservoirs contain gas that either is dissolved in oil or form a cap above the oil. It is called the associated gas. The production of oil in such cases leads to flaring the gas in the atmosphere in order to extract the oil underneath. Most of the gas reserves in the Gulf are associated, which means the production of oil is correlated to the production of gas. When Saudi Arabia cut back oil production to remain within the OPEC quota, an unintended consequence of the measure was that it had to reduce the flows for domestic power generation and industrial feedstock. 'The Kingdom may even have to think the unthinkable and consider importing gas in the future,' according to a Saudi economist.[12] The flared gas is a major pollutant; it increases global warming and is an enormous waste of a valuable resource. Since 1870, at least 230 tcf (6.5 tcm) of natural gas has been flared worldwide.[13] The flared gas is considered the wasted gas. The reinjection of gas into the oil reservoirs to enhance the oil recovery is now increasingly practised.

The stranded, orphaned or untouched gas is unusable gas. It stays underground rather than thrown out into the atmosphere. This gas could be physically stranded in the case where it is trapped beneath too deep an obstruction. It could be financially stranded where the transport to the market is too costly. The energy companies know where substantial oil and gas resources lie. They are not reserves because they are not shown to be producible under economic constraints, but a sufficient increase in price would instantly make some known resources become reserves.[14] Approximately 40 per cent of the world's natural gas reserves are classified as stranded gas. It is estimated that there is currently 3,000 tcf (85 tcm) of stranded gas across the globe.[15] For example, on the Alaskan North Slope some 840 bcm – 30 tcf – of gas has long been known. But that gas cannot be marketed because the costs of shipping the gas to Japan or California are higher than the prices in those markets. Similarly, large fields are known in Yakutia and in the waters around the Yamal Peninsula in Russia. There, too, distance dominates, and the gas resources remain unmarketed.[16] The future may yet see them coming on ground in changed market and improved transport circumstances.

Geographical concentration/spread

Since transport is the most important issue in gas trade, geography is of utmost importance. Gas geography, in broad brush, looks like this. Europe (except the Netherlands) and India are gasless regions surrounded by gas-rich regions. Russia has two big consumers on either side – Europe on the west and China on the east. Ukraine is perpetually subjected to interference due to its location on the energy map. Landlocked Central Asia is dependent on the access to the consumers.

As per the BP estimates quoted in Table 1.1, Iran, Russia and Qatar are the three largest holders of natural gas stocks: Iran at 33.8 tcm, Russia at 31.3 tcm and Qatar at 24.7 tcm. A cursory look at the map would bring out the proximity of the three gas-rich states, which together hold nearly 90 tcm of gas, making up nearly half of the world reserves. Iran and Qatar share one single gas field that spans 9,700 sq. km and contains non-associated gas. The Qataris call their share of it North Dome or North field; the Iranians call their share South Pars. The Qatari North field contains about 910 tcf (26 tcm), which accounts for 14 per cent of the worldwide natural gas reserves. The South Pars field, a geological extension of the North field, contains an estimated 280 tcf (8 tcm) of natural gas. Thus, this single accumulation contains about 20 per cent of the world's natural gas reserves.[17]

Russia and Iran share their offshore reserves in the Caspian Sea, where Azerbaijan, Kazakhstan, Turkmenistan and Uzbekistan also have considerable stocks of gas. Azerbaijan has 0.9 tcm, Kazakhstan has 1.5 tcm, Turkmenistan has 17.5 tcm and Uzbekistan has 1.1 tcm. Together, the Caspian littoral states have reserves totalling nearly 111 tcm of gas amounting to a substantial concentration of the commodity. The gas reserves, from this perspective, can be seen as a continuum from the Gulf to the Caspian.[18] The fresh discovery of gas in the eastern Mediterranean (also called the Levant basin) can be seen as a further geographical extension. In short, the gas reserves are more dispersed throughout the globe compared to the oil reserves which are overwhelmingly concentrated in West Asia.

The US Government Energy Information Administration (EIA) gives a detailed estimate of the proved and probable reserves in the Caspian basin. There are 292 tcf (8.2 tcm) of proved and probable reserves. Offshore fields account for 36 per cent of natural gas – 106 tcf (or 3 tcm), most of which are in the southern part of the Caspian Sea. In addition, the U.S. Geological Survey (USGS) estimates another 243 tcf (or 7 tcm) of natural gas in as yet undiscovered, technically recoverable resources. The large amount and the dispersed nature of Caspian natural gas reserves suggest the possibility of significant future growth in production. The EIA estimates that the Caspian Sea region produced 2.8 tcf (or 0.8 tcm) of natural gas in 2012, which amounted to around 3.4 per cent of the total world supply in 2012. A large portion of it was reinjected back into fields or flared. Azerbaijan became an important regional natural gas producer with the start of production in the Shah Deniz field in 2006. Other prospects for natural gas production growth include Russia's North Caucasus region, which has the bulk of the Caspian Sea region's onshore natural gas reserves and Turkmenistan's Galkinish field, which a 2009 audit suggested may be the world's fourth-largest natural gas field.[19]

The same source also gives a detailed estimate of recently discovered gas assets of the Levant basin in the eastern Mediterranean consisting of Cyprus, Israel, Jordan, Lebanon, Syria and the Palestinian territories. Of significant natural gas fields discovered over the past decade, almost all have been in the Levant basin. Most are in Israel's territorial waters, although there are significant discoveries in Cyprus and the Palestinian territories as well. The USGS estimates that the Levant basin has a combined proved natural gas reserves of 18.2 tcf (0.5 tcm) and mean probable undiscovered natural gas resources of 122 tcf (3.5 tcm). The USGS also estimates the mean probable undiscovered resources of natural gas liquids (NGLs) at 3.1 billion barrels. While the total combined proved reserves of natural gas in the Levant basin are insignificant relative to global levels, recent discoveries in the offshore waters of Cyprus and Israel in the Levant basin pushed the estimated recoverable natural gas resources in the area to over 38 tcf (1 tcm).[20]

The adjacent Nile Delta basin also in the eastern Mediterranean is estimated to contain 223 tcf (6.3 tcm) of undiscovered, technically recoverable natural gas.[21] The Levant basin gas is a late arrival on the scene, and exploration is still under way. Extraction and marketing will have to wait even longer. Under the best-case scenario, the Israeli gas is scheduled to arrive to the Israeli market by the end of 2017 and to the Asian market via a floating LNG terminal no earlier than 2020.[22]

Gas in the energy basket

Gas versus coal

Increasingly, gas is referred to in the expressions like the clean fuel, bridge fuel,[23] golden age of gas, dash for gas and stepping on the gas. They suggest a clear preference for gas at present and in the coming years. The ranking of gas among the energy sources and the share of gas in the energy mix are constantly on the rise. According to the International Energy Agency, natural gas will continue to increase its share of the global energy mix, growing at 2.4 per cent per year until 2018.[24] Gas has largely replaced coal as the fuel to generate electric power and is expected to replace it as the most important source of electricity by 2030. Electrical utilities like Germany's RWE are considering if it would be economically viable to use coal to generate electricity.[25] A UNDP/World Bank study terms this trend a 'dash for gas', which is reinforced in many areas by a combination of gas-sector reform, creating gas-to-gas competition, and electricity-sector reform, leading to strong demand for combined-cycle gas turbine (CCGT)[26] generation and concerns about the environmental damage caused by the consumption of other hydrocarbons.[27] Over the next 30 years, world demand for gas is expected to double, surpassing coal as the world's second-largest energy source.[28]

There is a strong counter-argument on this point. There is, in fact, a strong campaign against gas: Stop the Dirty Fuel Campaign. Debunking the claim of gas being a clean fuel, the campaign calls it the 'dirty clean fuel', as the gas is prone to leaking from the pipelines, well heads, and the nooks and crannies of the processing and storage facilities. Accounting for the methane leakage throughout the supply chain, gas is worse for the climate than coal, it out-pollutes coal and it should be called an exit door rather than a bridge fuel to the low carbon energy future.[29] No fossil fuel

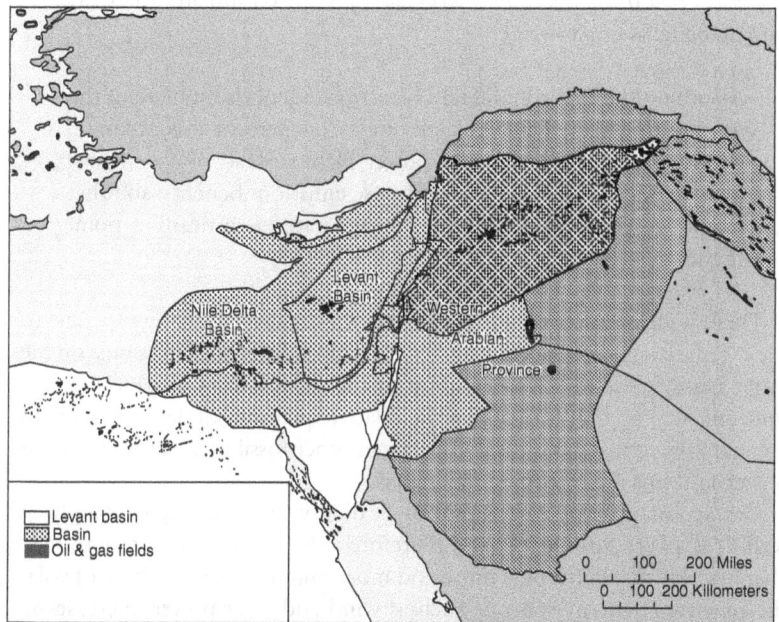

Figure 1.2 Eastern Mediterranean basins.

Source: Based on 'Eastern Mediterranean Region', *Energy Information Administration*, 15 August 2013. http://www.eia.gov/countries/analysisbriefs/Eastern_Mediterra nean/eastern-mediterranean.pdf. Accessed on 21 October 2013.

Disclaimer: Map not to scale.

is without its polluting minus points. It is the comparative adverse impact on the environment that should be the standard for grading the various sources of energy. That gas is worse than coal on such a criterion is difficult to subscribe to. A differing opinion has its own space in any situation, and the one above needs to be noted as well.

Gas versus oil

Comparing gas to oil, there are powerful arguments on both sides. On the environmental consideration, gas wins hands down. Substitution of gas for high-sulphur coal and oil would lead to an almost zero emission of sulphur dioxide. Because methane molecules contain more hydrogen and less carbon than those of other organic fuels, natural gas combustion produces less carbon dioxide and less 'greenhouse' atmospheric warming effect per unit of useful energy than use of coal, oil or biomass.[30] A study published in the

Proceedings of the US National Academy of Sciences estimates the sustained climate benefits of gas usage:

> Assuming the Environmental Protection Agency's estimate of the gas industry's methane leakage rate – 2.4 percent – is accurate, choosing to build a new gas power plant instead of a new coal plant produces immediate greenhouse emissions benefits, and that would be the case even if the leakage rate were nearly a point higher.[31]

There is an alternative finding from the International Energy Agency: to the extent that gas displaces oil and coal, it drives down the emissions; on the other hand, gas also displaces some of the nuclear power, which pushes the emissions up.[32] It is silent on the relative quantum of displacement. In comparative terms, gas is located between other fossil fuels and the nuclear power in terms of environmental impact.

Gas scores over oil on counts of flexibility and security of supply as well. It is a very flexible fuel and therefore able to respond quickly to load changes, which will become more and more important as the share of volatile renewable energy sources, such as wind and solar power, increases in the overall energy mix.[33] There are two arguments in favour of the gas on account of security of supply. One, security is assured precisely because of the presence of oil as an alternative and as a fallback option. Two, a sudden drop or increase/decrease of gas supply could lead to major safety hazard and, therefore, not be casually resorted to by either party.

On the other hand, oil scores over gas in versatility, cost, infrastructure and time. Oil has no competitor in the transport sector, whereas gas faces many alternatives like coal, oil and nuclear energy for generating electricity. Gas has a much lower density than oil – at standard temperature and pressure, it contains much less energy than the same volume of oil. That translates into much higher costs for storage and transport. Security of gas storage requires it to be fully contained at all time to prevent it from mixing with air or escaping. To be economical, the gas density needs to be increased either by storing it at very high pressure or by liquefying it at low enough temperature. Oil has a much more robust and interconnected infrastructure than gas; damage to oil-receiving ports, storage sites and refineries can be replaced by increased transportation or from other installations. A damaged gas facility or pipeline, on the other hand, would need to be isolated, and the market would need to be well connected to other potential sources by a web of pipelines.[34] Compared to oil, the time frame for a gas project to become functional is long – even in the absence of any technological or financial obstructions.

Notwithstanding relative advantages of coal and oil on some counts, the future trend towards gas is unmistakable. Corporations like Exxon, BP and Shell, which have seen themselves primarily as oil producers for generations, are now investing billions in the gas industry.[35]

Gas markets

Over 80 per cent of gas consumed in the world is produced locally. The natural gas, to that extent, is a national commodity, rather than an international exchange.[36] Majority of gas supplied (approximately 70 per cent) is consumed domestically in gas-producing states. The majority of the internationally traded gas is supplied within regions, and gas trade between regions across the globe is still limited.[37] It is easier to export oil rather than gas. Gas, therefore, is used domestically and oil is exported. It is a normal trade practice of the oil- and gas-producing countries. This may not always be possible, as the domestic demand may vary, the energy production profile may be different and the export opportunities may be limited.

Next to the domestic usage comes the regional trade. The regional clusters have mostly been enduring; for example most of the imported gas in the US comes from Canada; Europe imports most of its gas from Russia and Algeria; and the consumers in Asia like China, India, South Korea and Japan source it from Qatar, Australia, Brunei, Indonesia and Malaysia. In East Asia, the gas trade has been an integrative factor, with more than half of Southeast Asian gas exported to the three Northeast Asian consumers: Japan, Taiwan and South Korea.[38]

A typical regional gas market, in this instance, Europe, works like this: the big European players, like Gazprom in Russia and Statoil in Norway, exploit their reserves and then transport the gas through thousands of kilometres of pipelines to deliver it to the border of Germany or other European nations. From there, distributors like Ruhrgas or Wintershall feed the gas into their networks and sell it to municipal utilities or industrial customers. It is a profitable business for everyone involved. Long-term agreements are in place with terms of up to 40 years, and they are based on the so-called gas–oil price link, which means that gas prices follow oil prices, only with a few months' delay. The distributors add a healthy margin of up to 30 per cent for the distribution, storage and sale of the gas. For a company like Ruhrgas, this means that with its roughly €2 billion in annual profits, it is the most important subsidiary within the E.on group.[39]

It is said that the oil contracts are commercial deals, but the gas contracts are similar to a marriage. The relationship between the host country and the gas companies could be decades old even before the gas is brought on the ground. Compared to oil deals, the relationship is likely to be much stronger because of the restrictions on delivery mechanisms. Gas is not yet a fungible source of energy, and the pipeline delivery ties the producers and the consumers in a long-term interdependence-reinforcing regional alliances. And since the gas reserves in the world are more dispersed than oil, it is easier for the producers and the consumers to work within a regional context. A beneficial fallout of this is that the shortage in one sector does not affect the larger trade.

The segregated regional markets have gradually started to coalesce towards an integrated global market. The trans-Atlantic liquefied natural gas (LNG) trains from Australia to Spain, from Qatar to Europe and potentially Asia-Pacific, have made the network of the gas tighter and wider. Three factors have helped the process along: technological development; lowering of the costs of liquefaction, transportation and re-gasification; and liberalisation of the markets in an increasingly globalised world. As a consequence, for the first time, something resembling competition has developed in the gas industry. The volumes being traded on the spot markets have got bigger and bigger. New competitors could buy up gas at favourable terms, which has benefitted consumers, who could choose from among an average of 31 gas providers, as compared with only 8 traditional providers.[40] The globalising market would have potential implications for pricing. New sources of gas would bring greater diversity to the global supply.

A joint study by the Harvard University and the James Baker Institute concluded that a series of developments like increasing demand, technological advances, cost reduction in producing and delivering LNG to the market and market liberalisation were spurring the integration of gas market. Such market interconnections would have major ramifications for both large gas consumers and producers.[41] The International Energy Agency singles out the LNG as the glue that would increasingly bind the three regional markets of the US, Europe and Asia-Pacific.[42]

An alternative scenario depicts a continuation of regional markets with greater interaction among them. A multidisciplinary study done by the Massachusetts Institute of Technology concludes that even in the case of more integrated markets evolving, the regional markets will be distinct with respect to supply patterns, pricing, contract structures and market regulation, which will lead to rising trade among the current regional markets.[43] The global market, according to this study, will be a conglomeration of regional entities. Changes in the cast of major LNG suppliers would create new linkages between regional markets, notably between those of North America and the Asia-Pacific, narrowing to a degree the wide regional gas price differentials that exist today. The convergence between regional gas prices would push the prices down and gas demand up (by 107 bcm in 2035) and lower the gas import bills.[44]

The discovery of gas in the eastern Mediterranean and shale gas in the US is perceived to encourage a form of 'energy regionalism'[45] and swing the markets back to the regions. This trend is likely to reduce the attractiveness of large infrastructural projects. Whether this prognosis comes to pass, only will time tell. The reasons advanced for it materialising do not seem very convincing. The gas in the eastern Mediterranean will take a few years to come on line, and even when the US gas does reach Europe, it will happen in a more globalised world than what exists today. The loosening of regional boundaries and the connections between the regions, once initiated, is hard to rein in again.

And then there are perceptions of risks involved in the gas trading. There is a great concern regarding the situation in the producing countries, where laws and institutions are weak; governments challenge contracts that companies considered ironclad; fiscal regimes change unpredictably; environmental fines are imposed somewhat whimsically; ethnic, clan and various civil conflicts occur; corruption is rampant; and regimes are ever more authoritarian.[46] There are concerns that go beyond the producing countries to take in the entire gas trade. According to this perception, gas trade requires securing supplies that originate in, cross and arrive in countries where contracts are difficult to enforce, regulatory systems are immature, and investors have been wary in deploying capital. 'Risky countries', as they are termed, are likely to play a pivotal role in the trade. They include large potential gas suppliers (Russia, Iran, Saudi Arabia, Qatar, Nigeria, Algeria, Indonesia, Venezuela, Trinidad and Bolivia), key potential transit countries (Ukraine, Turkey, Pakistan and Chile) and new centres for potential gas demand (the rapidly growing economies of China, India and Korea).[47]

This seems to be an extraordinary perception. What is the most objectionable about the perception is its inclusiveness. Not very many countries – apart from the US, Canada and West Europe – are left out of the label 'risky'. A closer look would reveal that politics influences 'non-risky' countries as well. The US government sanctions on the gas-rich countries like Iran make little commercial sense as they prohibit the US companies to trade with or invest in the Iranian gas-sector development. For American companies operating overseas, a second type of political risk arises from pressure from a US administration or from the US Congress. This may take the form of pushing companies to pursue pipeline projects that may have strategic, but no commercial, appeal.[48] The gas companies were very reluctant to go along with the US administration–preferred Baku–Tbilisi–Ceyhan (BTC) and Baku–Tbilisi–Erzurum (BTE) pipelines. Both had only one raison d'être: to keep Russia and Iran out of the global energy loop.

The extractive industry – be it gas, oil or metal – is never a simple give-and-take. It does involve people and politics through the entire chain of the trade. The people in the originating countries feel robbed of their resources, the people in the transit countries feel cheated by the low rents for the transit facility and that they are under a constant threat of disruption by the originating country and the people in the consuming countries complain about the higher price they have to pay than the others.

Gas market is a capital-intensive market par excellence. Before the geologist even comes into the picture or the engineer assesses the feasibility of extraction, it is the investor who has to commit the finances to make the project doable. The vast array of infrastructure requiring enormous expenses would include the onshore and offshore extracting platforms, pipelines, machinery and equipments; LNG liquefaction, transport and re-gasification plants; and several other overheads. A typical gas export deal would require devoting more than half of the capital investment in the exporting country. Once constructed, the investment is immobile. The expense incurred would begin to bring returns only after years or even decades. In the meanwhile, the investment would remain immobile in the host country – hostage to the hosts.

And once the capital is sunk into developing gas fields and constructing export infrastructure, the investor loses the position of power that he held on account of scarce capital and advanced technology. The leverage for setting price and other terms of trade passes to the host country government and the gas-buying company/government. In a typical situation, the producer has to manage both the backward and forward linkages of the chain. In this situation, the investor deploys his own defensive strategies. Joint ventures, partnerships and under-the-table sweeteners are common tricks of the trade. He may also choose to employ technologies that require special expertise to operate, rendering his investments unattractive for expropriation, just as retreating armies poison wells and burn depots to spoil the rents of territorial advance.[49]

Gas-Exporting Countries Forum versus Gas Buyers' Club?

The four major gas producers – Russia, Iran, Turkmenistan and Qatar – do have common interests and concerns. Their informal consultations resulted in the establishment of the Gas Exporting Countries Forum (GECF) in May 2001 in Tehran. It took a few years before it was institutionalised at the ministerial-level meeting in Moscow in December 2008. Algeria, Bolivia, Egypt, Equatorial Guinea, Iran, Libya, Nigeria, Qatar, Russia, Trinidad and Tobago, the United Arab Emirates and Venezuela are members of the forum, and Doha is the forum headquarters. Kazakhstan, Iraq, the Netherlands, Norway and Oman are its observer members. A conspicuous absence from the lists of members and observers is Turkmenistan. Apart from ministerial meetings, three summits of the forum have been held to date – the first in Doha in November 2011, the second in Moscow in July 2013 and the third in Tehran on 23 November 2015.

The Tehran Summit underscored the need to facilitate stronger cooperation over a series of industry issues, including the transfer of expertise and pricing mechanisms. The significance of the event went far beyond the issues related to natural gas. It was attended by heads of state from nine member countries, including presidents of Iran, Russia, Venezuela, Iraq, Bolivia, Equatorial Guinea, Nigeria, Turkmenistan as well as the Algerian prime minister. Two of the attendees deserve to be noted in particular. The Turkmen president was invited as a special guest even though Turkmenistan is not a member of the forum. The Russian president Vladimir Putin visited the country after a gap of eight years and had a meeting with the Iranian supreme leader Ayatollah Ali Khamenei that was billed as the 'Meeting of the Titans'. Positive externalities to the forum or to the global gas trade in general, if any, remain to be seen.

As the name itself suggests, it is a forum and not a cartel. The overarching goal it has set for itself is to support the sovereign rights of member countries over their natural gas resources and their abilities to independently plan and manage the sustainable, efficient and environmentally conscious development, use and conservation of natural gas resources for the benefit of their people. It seeks to contribute towards 'development of transparent, efficient and competitive regional and global gas markets able to withstand current and future risks and challenges'.[50] Cross investments and technological exchanges are sought to be promoted. Overall, the GECF is projected to be a group that has common interests and therefore a common platform.

There are several reasons that the GECF cannot, and does not even intend to, become a cartel. First, the high cost of transporting gas and competition from other fuels has placed limits on the evolution of GECF into a gas cartel, popularly referred to as gas OPEC. Second, given the high costs of trading natural gas, prices are usually locked in long-term bilateral contracts (20 years or longer) with a take-or-pay clause. This means that buyers are obligated to pay for the gas whether they take it or not. This rigidity of gas pricing leaves little room for GECF to try to influence gas production and prices.[51] Third, piped gas cannot be redirected anywhere beyond where the pipe leads. The supplier intending to cut off supplies would have no alternative buyer and would be unable to recoup lost revenue.[52] A pipeline is a double-edged sword hurting both.

Apart from the market problems, there are four issues related to the membership. One, the Netherlands and Norway, located in gas-consuming

Europe and currently observer members, may not like conflict with consumer countries on issues of price and production. Canada, till recently, was fully committed to the US market and has not become a GECF member. A consensus among its members will be unlikely in this grouping. Two, it may not be possible to coordinate the policies of the forum with the OPEC on all issues. A forum member country need not be an OPEC member country, in which case the ensuing clash of interest between the members may spill over into both the groups, weakening them both. Three, there is an interesting phenomenon of the emergence of the importer/exporter. Countries like the US, Canada and Iran with plentiful gas resources and production need to import gas to supply certain areas due to transport bottleneck between producing and consuming regions within their own territories. Indonesia, Malaysia, the United Arab Emirates and Oman also need to import gas to meet the export commitments and the growing domestic demands. Egypt, a traditional exporter, is in the process of diverting the gas supplies intended for the LNG export to the domestic market and seeking to import gas from Israel.[53]

And finally, there is a structural problem. A cartel requires a swing producer to monitor the price/production on the ground adhere to the agreed-upon formula. Saudi Arabia played that role within the OPEC since its inception till very recently. The role of a swing producer will be much more costly for a gas producer because of the much higher fixed costs associated with gas projects. For gas, the swing producer will have to maintain excess liquefaction and LNG tanker capacity (or gas pipeline capacity as well as extra production and storage capacity). Given the costs, such a role may not be attractive and consequently go unfilled.[54] There is speculation whether the US would emerge as the swing producer, as its shale gas production picks up. Would it be willing to?

The most fundamental issue is that Russia is an uncontested leader in gas export, and Putin is considered the best energy expert among the world leaders.[55] Russia would, therefore, play its own game unfettered by extraneous considerations, although shackled by sanctions at present. Iran is preoccupied with coming out of the web of sanctions and may not be looking at playing the role. Qatar is fully into the LNG trade, far away from the game of the gas pipelines at present. With these three top owners and exporters of gas going their own ways, a cartel simply is out of the question.

At the other end of the gas trade, the gas buyers have attempted to form a group of their own. The rising demand in the Asian markets on the one hand, and a shrinking supply due to the shutdown of Japanese and Korean nuclear plants and a shift towards cleaner gas in China on the other, has created a drastic gas shortage in Asia. The shale gas exports from the US, simultaneously, have remained restricted to select countries. Together, these factors have set the global trend towards the gas market turning into a sellers' market.

In these circumstances, India took the initiative in reaching out to other major gas buyers like China, Japan, South Korea and Taiwan. Together, these five countries account for roughly two-thirds of the LNG traded internationally. They have nursed a grudge at the cost discrepancies they face – paying at least four times as much as the 'Henry Hub'[56] price, the North American benchmark. Some of it is explained by transport and shipping costs, but they complain about both structural problems and limited transparency.[57] According to an energy analyst, Asia's system of buying gas but paying for oil – akin to buying water but paying for champagne – imposes a hidden tax that prevents natural gas from competing not only against coal in power generation but also against oil in the transportation sector. The inability of natural gas to compete against coal and oil denies Asia its most effective mechanism to combat air pollution – switching to a cleaner fuel.[58]

At a conference jointly organised by the Indian Ministry of External Affairs and the Federation of Indian Chambers of Commerce and Industries in New Delhi in September 2013, the chairman of the Gas Authority of India Limited (GAIL), B. C. Tripathy, announced that all the major buyers of LNG would formally launch a Gas Buyers Club on 3 December.[59] Accordingly, an Asia Gas Partnership Summit was held in New Delhi on 3–4 December to deliberate the issue. The buyers' agenda was expected to include reduction of the cost of the long-term LNG contracts, delinking contracts from oil prices and eliminating the clauses that restrict the destination of shipments and prevent them from selling the excess cargoes. The Asian buyers were hopeful of betting on buying the LNG shipments from the US, which has started relaxing restrictions on gas exports.[60] There was an expectation that the meeting may mark a step towards the formation of the Gas Buyers Club, announced earlier by Tripathy.

As it happened, the summit did not conclude on the announcement of a formal launch of the Gas Buyers Club. The explanation for this could be that they had set themselves an extremely difficult task. The very nature of gas contracts would preclude renegotiating an existing contract, which typically runs for 20 years and more. In the case of the LNG contracts, the buyers would have to manage logistics and negotiate harder with shipping companies, in addition. And demand-and-supply needs would have to be closely monitored since the buyers do not have cross-country pipeline network that connects with each other.[61]

Thus, the date set for the launch of the club came and went. It was only towards the end of January that GAIL and Japan's Chubu Electric Power Company came together to jointly buy liquid gas with a view to leveraging their combined purchasing capacity to beat down prices. B. C. Tripathy announced that the two would sign a Memorandum of Understanding on the joint purchase.[62] It was still nowhere close to the launch of the Gas Buyers Club. Even the bilateral deal with Chubu was a work in progress, at that stage.

Tired of a lack of progress on the government-led initiative, a gas industry conference was organised in Seoul, South Korea, by some Asian gas and power companies in April 2014. The issue of the Gas Buyers Club was not even on the agenda, and Japanese companies were not even represented by the senior management.[63]

Nothing has been heard of the Gas Buyers Club since then.

Future of/with shale

The discovery of gas was accidental. Mysterious eternal fires scared people at first. Soon, the fires were put to use. It was easy, like squeezing a sponge. Progressively, the extraction had to be done at deeper and deeper layers. The offshore discoveries led to extraction under water, under deeper and deeper waters and further and further from land. It has been a journey from squeezing the sponge to breaking the mountains and churning the oceans. Fracking is yet another step in humankind's search for energy.

Hydraulic fracturing or 'fracking' is the process of injecting sand, water and chemicals into shale rocks to crack them open and release the hydrocarbons trapped inside. It is a sort of controlled earthquake. Today drilling companies can drive their wells thousands of metres beneath the surface, divert the drill heads and even continue drilling horizontally. The high-tech tunnelling machines can reach a target location even when it is 8 km away. Once the target has been reached, 'frack trucks' bring giant 2,400-horsepower pumps to the site, where about a dozen of them are connected. They force a fluid mixture of millions of litres of water, sand and chemicals into the gas deposit at a great pressure. The process produces enough pressure underground to fracture the rock. This creates fine cracks, some of them hundreds of metres long. The fluid is pumped out of the well and the gas escapes, powerfully at first, and then more slowly for several months, until the pressure is so low that the fracking procedure has to be repeated.[64]

Shale gas is nothing new. Geologists have long known that trillions of cubic metres of natural gas are trapped in dense underground rocks. The first commercial natural gas well drilled in the US – in Fredonia, New York, in 1821 – was in shale.[65] Devon Energy, already a successful oil and gas development firm, seized on an innovation in natural gas in the early 2000s. Working in the Barnett shale of Texas, it combined two technologies that had long been used separately in the oil and gas industry

– horizontal drilling and hydraulic fracturing – to stunning effect, massively boosting production from shale gas resources and turning vast expanses once thought uneconomic into ripe development targets.[66] In more recent times, the production of shale gas in the US has been going on for decades. The reason that it suddenly grew by 71 per cent in the year 2008 was that in previous years exploration technologies allowing the geologists to find the reserve figures had been grossly underestimated.[67]

The EIA estimates that shale will add approximately 47 per cent to the 15,587 tcf (441 tcm) of proven and unproven non-shale technically recoverable natural gas resources.[68] In many parts of the world, geologists are now testing the ground for natural gas trapped in shale that was largely unreachable in the past. Advanced Resources International, an energy consultancy, has estimated that the sheer volume of Chinese shale gas resources exceeds that of US resources.[69] The estimated reserves in China are 36.1 tcm, followed by the US at 24.4 tcm, South America and Africa. In Europe, the largest estimated gas reserves are in Poland, 5.3 tcm; France, 5.1 tcm; Norway, 2.4 tcm; Ukraine, 1.2 tcm; and Sweden, 1.2 tcm. No shale gas reserve estimates exist for leading gas exporters in Russia, Iran and the Gulf, perhaps because their vast conventional gas reserves render the point moot.[70]

The Americans are the uncontested winners in the shale game. For the Europeans, the shale revolution is largely positive. By 2020, shale gas in the form of LNG is likely to begin arriving in Europe in significant quantities, and there is also the prospect of some domestic shale gas becoming available.[71] The domestic production of shale gas in Europe is not likely to be substantial. The producers there would need to operate within the densely populated and highly environmentally conscious EU member states. A strict legal system, especially with regard to property rights and water management issues, would further restrict the process.[72] In addition, many European countries lack the oil-and-gas service sector, production infrastructure, relatively easy access to land and a permissive regulatory environment that have eased shale gas development in North America.[73]

Shale has ardent supporters and vocal opponents. The critics of the shale revolution point to three drawbacks: one, the sharp inflection points in the shale gas wells result from horizontal drilling and hydraulic fracturing technologies. Production peaks for a year or two, but then the initial flow peters out. Overall lifespan of shale wells in Texas is about eight years. Drilling company must continuously invest in the new wells or re-frack the old ones. In comparison, conventional, vertically drilled wells demonstrate more stable output for 20–30 years.[74]

Two, the production has led to popular protests in many cities across the US. People are worried that the highly contaminated sludge containing benzene, xylene and toluene shoots back on the ground and pollutes underground aquifers.[75] Small earthquakes are seen to be triggered by the fracking activities in the area. The protesters complain that the climate change; damage to the roads, houses and other utilities in the neighbourhood; noise; and heavy traffic lead to a sharp deterioration in the community life. They demand fair compensation for the lost land and are worried about the terror attacks. The protests have remained localised and uncoordinated in the nature of 'Not in my backyard' or NIMBY. The environmental advocates would prefer a battle cry around 'Not on our planet' or NOOP, instead.[76]

Three, there are reports that reveal how investment bankers promoted shale bubble in order to profit from a short-lived energy boom. Just like the subprime mortgage crisis, the shale may turn out to be a deliberate creation of financial bubble. And like the Gold Rush of the nineteenth century, the shale rush may turn out to be a limited market opportunity. A variation of the same theme sees the shale as an inflated subject created by the rivals of Russia and as nothing more than a soap bubble inflated to knock down the prices and shift the price-formation process from the currently accepted long-term contracts to short-term contracts.[77] The Russians are convinced that their gas is the target of the shale bubble.

The shale revolution has raised a plethora of questions. It will not remain an American phenomenon for long. Others will catch up, sooner or later. The short time span of shale gas wells may, in the process, continue from location to location, assuaging the fears of a dream that would vanish with the first ray of the sun. The process of fracking may improve with better technology and newer devices.[78] As severe problems associated with it are solved and as the benefits become apparent,[79] the protests may tone down.

Alternatively, a major mishap may bring the entire shale revolution to a screeching halt. No other state, in the eventuality, may want to go the American way. It is too early a stage to make prognosis on the shale future. It is not beyond the realm of possibility that in a few years' time the world may be pondering 'world without fracking' on the lines of 'world without the weapons of mass destruction'.

All this is future. At present, shale development has made the US self-sufficient in its energy requirement and soon will make it a credible exporter.[80] Before the shale, the US was building terminals to import gas. Now, the five terminals lie idle as plans are made to turn some of them into platforms that can allow the gas exports.[81] Europe will be a major beneficiary of the shale development in the US. Its overwhelming dependency on Russian gas imports would be alleviated to a large extent.

Shale is an extremely negative development for Russia. Its economy is based on the exports of oil and gas, and it has invested huge sums to build the pipelines to Europe. A Russian energy expert rues the fact that while Russia was busy building pipelines and a system of oil and gas extraction in complex climatic conditions, Americans achieved a technological break-through and organised consistent and profitable production of shale gas.[82]

Recapitulation and leads

Natural gas consists mainly of methane, which is one atom of carbon and four atoms of hydrogen (CH_4), the simplest hydrocarbon. Its origin could be fossil, thermogenic, mineral or biogenic. Its history goes back thousands of years, but its commercial production and use are of recent origin. The great period of gas discovery began in the mid-1950s, peaked in 1960s and extended throughout 1970s. The gas discovery begins with the geologist who locates the gas-rich terrain, nowadays with the help of seismic devices and satellite imagery that indicate the presence of gas below surface. The global total proved reserves of gas are estimated to be 185.7 tcm and unproven technically recoverable resources to be 441 tcm. The EIA estimates that shale will add approximately 47 per cent to the natural gas resources. As new discoveries come on ground, as new technologies become available and as better methods of transport are devised, larger volumes of gas will become exploitable. On the other hand, with other variables remaining constant, the gas consumption will deplete the reserves and drive the estimates downwards. The gas that is flared into the atmosphere is wasted gas. The gas that remains underground because it is too remote from the market or because it is too prohibitive in cost is termed stranded, untouched or orphaned gas. It is estimated that there is currently 85 tcm of stranded gas across the globe. Gas is dispersed throughout the planet, though there is a rough linkage from Qatar to Iran to Russia and now the Mediterranean. The gas market is regional in nature due to transport consideration; it is evolving towards a global market due to technological innovations and market forces. Neither the gas sellers nor the gas buyers have been able to come together to form an exporters' or a buyers' cartel so far. The chapter opened with the quotes that ranked the natural gas as the best among the coal, oil and nuclear fuels, and forecast its continuing and expanding utility in coming years as the best energy source. It should close by validating the quotes and adding that the shale gas is likely to reinforce the position of gas even more – at least for a few decades.

Notes

1 Arlon R. Tussing and Bob Tippee, *The Natural Gas Industry: Evolution, Structure and Economics* (PennWell Books, Tulsa, 1995), p. 47.

2 Maria van der Hoeven, Executive Director, International Energy Agency, 20 June 2013. http://iea.org/newsroomandevents/pressreleases/2013/june/name,39014,en.html. Accessed on 4 July 2014.

3 Rebecca L. Busby, ed., *Natural Gas in Non-Technical Language* (PennWell, Tulsa, 1999), pp. 2–3.

4 Paul Gretton-Watson, 'Energy: Wasted at the Wellhead', *Bulletin of the Atomic Scientists*, Vol. 58, no. 5, September 2002, pp. 22–23.

5 Tussing and Tippee, n. 1, pp. 25–26.

6 'About Gas', *Eurogas*, 2014. http://www.eurogas.org/about-gas/. Accessed 10 October 2013.

7 Trapped below the permanently frozen ground (permafrost), Urengoy is estimated to contain 8 trillion cubic metres (tcm) of gas.

8 Busby, n. 3, p. 5.

9 David Hawdon, 'Market Competitiveness: The Economics of the Natural Gas Market and Its Competitiveness', in Paul Stevens, ed., *International Gas: Prospects and Trends* (Macmillan, Houndmills, 1986), pp. 14–33.

10 *BP Statistical Review of World Energy*, June 2014. http://www.bp.com/ content/dam/bp/pdf/Energy-economics/statistical-review-2014/BP-statistical-review-of-world-energy-2014-full-report.pdf. Accessed on 15 October 2014.

11 Compare, for example, the BP aggregate gas reserves figures of the Caspian basin littoral states with the Energy Information Administration (EIA) figures that are quoted a little later. The latter estimates include not only the onshore and offshore energy assets in and around the Caspian basin but take in even the probable reserves in addition to the proved ones.

12 Quoted in Andrew England, 'Overlooked Resource', *Financial Times* (London), 26 May 2008. See also *Petroleum Intelligence Weekly*, Vol. 48, no. 9, 2 March 2009.

13 Gretton-Watson, n. 4.

14 Ben W. Ebenback, *Energy Resources: Availability, Use and Impact* (PennWell, Tulsa, 1995).

15 http://www.energtek.com/t/1029-stranded-gas. Accessed on 12 September 2014.

16 Thomas R. Stauffer, 'Caspian Fantasy: The Economics of Political Pipelines', *Brown Journal of World Affairs*, Vol. VII, no. 2, Summer–Fall 2000, pp. 63–78.

17 'North Field and South Pars Natural Gas Fields'. http://www.theenergy library.com/node/568. Accessed on 12 September 2014. There are differing estimates that put the number much higher.

18 There is a differing perception on the geographical spread/concentration that holds that there are two geographically separate regions of high gas reserves – Caspian and the Gulf – as opposed to oil where West Asia is of unparalleled importance. Amy M. Jaffe and Ronald Soligo, 'Market Structure in the New Gas Economy: Is Cartelisation Possible?', in David G. Victor, Amy M. Jaffe and Mark H. Hayes, eds., *Natural Gas and Geopolitics: From 1970 to 2040* (Cambridge University Press, Cambridge, 2006), p. 442.

19 'Oil and Natural Gas Production Is Growing in Caspian Sea Region', *Energy Information Administration*, 11 September 2013. http://www.eia. gov/todayinenergy/detail.cfm?id=12911. Accessed on 12 October 2014.

20 'Eastern Mediterranean Region', *Energy Information Administration*, 15 August 2013. http://www.eia.gov/countries/analysisbriefs/Eastern_ Mediterranean/eastern-mediterranean.pdf. Accessed on 21 October 2013.

21 'Technical Announcement: Nile Delta Natural Gas Potential Is Significant', *US Geological Survey*, 18 May 2010. http://www.usgs.gov/newsroom/article.asp?ID=2466#.VEn83LkcTIU. Accessed on 24 October 2014.

22 Gal Luft, 'Israel's Zero Gas Option: Take II', *Journal of Energy Security*, Spring 2014. http://ensec.org/index.php?option=com_content&view=article&id=533:israels-zero-gas-option-take-ii&catid=143:issuecontent&Itemid=435. Accessed on 1 October 2014.

23 The term implies reduction in greenhouse gas emissions that can bridge the world into a lower-carbon direction till the time cleaner energy solutions are invented and made available.

24 *International Energy Agency*. http://www.iea.org/topics/naturalgas/. Accessed on 25 February 2015.

25 Frank Dohmen, Alexander Jung and Jan Puhl, 'Stepping on the Gas: New Drilling Technologies Shake up Global Market', *Spiegel Online International*, 3 March 2011. http://www.spiegel.de/international/business/stepping-on-the-gas-new-drilling-technologies-shake-up-global-market-a-748573-2.html. Accessed on 14 September 2013.

26 The CCGT combines a gas-fired turbine with a steam turbine. The design uses a gas turbine to create electricity and then captures the resulting waste heat to create steam, which in turn drives a steam turbine, significantly increasing the system's power output without any increase in fuel.

27 'Cross-Border Oil and Gas Pipelines: Problems and Prospects', *Joint UNDP/World Bank Energy Sector Management Assistance Programme (ESMAP)*, June 2003, p. xiii. http://siteresources.worldbank.org/INTOGMC/Resources/crossborderoilandgaspipelines.pdf. Accessed on 14 September 2013.

28 Amy M. Jaffe, Mark H. Hayes and David G. Victor, 'Conclusions', in Victor, Jaffe and Hayes, eds., n. 18, p. 467.

29 Jeremy Deaton, '"The Dirty Clean Fuel": Why Natural Gas Out-Pollutes Coal', 5 November 2015. http://www.livescience.com/52715-natural-gas-not-as-clean-as-people-think.html. Accessed on 2 January 2016.

30 Tussing and Tippee, n. 1, p. 47.

31 Quoted in the *Washington Post*, 16 April 2012.

32 'Are We Entering a Golden Age of Gas?', *World Energy Outlook 2011*, *International Energy Agency*, p. 8.

33 Christopher Ross, 'EU Energy Policy in the Caspian Region and Central Asia', Unpublished Paper presented at the International Conference on *Energy, Transportation, and Economic Links in Eurasia: Emerging Partnerships*, New Delhi, 16–17 January 2012, p. 1.

34 'Natural Gas Market Review, 2007', *International Energy Agency*. http://www.iea.org/publications/freepublications/publication/gasmarket2007.pdf. Accessed on 14 October 2014.

35 Dohman, Jung and Puhl, n. 25.

36 Paul Stevens, ed., *International Gas: Prospects and Trends* (Macmillan, Houndmills, 1986), p. 124.

37 Brenda Shaffer, 'Natural Gas Supply Stability and Foreign Policy', *Energy Policy*, Vol. 56, 2013, pp. 114–125.

38 Robert A. Manning, 'The Asian Energy Market: A New Geopolitics?', in *Asian Energy Markets: Dynamics and Trends* (Emirates Center for Strategic Studies and Research, Abu Dhabi, 2004), p. 37.

39 Dohman, Jung and Puhl, n. 25.

40 Ibid.

41 'Geopolitics of Natural Gas' (A joint study by the James A. Baker III Institute for Public Policy's Energy Forum and Harvard University's Kennedy School), *Baker Institute Study*, no. 29, March 2005. http://www.bakerinstitute.org/publications/study_29.pdf. Accessed on 13 June 2012.

42 'Natural Gas Market Review 2006: Towards a Global Gas Market', *International Energy Agency*, 2006, p. 17.

43 'The Future of Natural Gas', *A Multidisciplinary MIT Study*, 2011. http://mitei.mit.edu/system/files/NaturalGas_ExecutiveSummary.pdf. Accessed on 3 March 2013.

44 'World Energy Outlook 2013 Factsheet: How Will Global Energy Markets Evolve to 2035?', *World Energy Outlook 2013*, 12 November 2013. http://www.worldenergyoutlook.org/media/weowebsite/factsheets/WEO2013_Factsheets.pdf. Accessed on 13 July 2013.

45 Ilektra Tsakalidou, 'The Southern European Corridor', *EU Institute of Security Studies, Issue Alert*, no. 21, July 2013. http://www.iss.europa.eu/uploads/media/Alert_Trans-Adriatic-_Pipeline.pdf. Accessed on 1 October 2014.

46 Maureen S. Crandall, *Energy, Economics and Politics in the Caspian Region: Dreams and Realities* (Praeger Security International, Westport, 2006), pp. 24–25. The author mainly focuses on the Caspian region, where a project's rate of return must be higher than normal to offset the risk. This risk premium alone has been estimated at around 10 per cent.

47 Mark H. Hayes and David G. Victor, 'Introduction to the Historical Case Studies: Research Questions, Methods, and Case Selection', in Victor, Jaffe and Hayes, eds., n. 18, p. 28.

48 Crandall, n. 46, p. 25.

49 European buyers of the Soviet gas in the 1980s lent capital to the USSR for construction of the massive inter-continental gas pipelines. However, the Europeans were quite sure that they would receive repayment for the loans, as the bulk of the capital they provided was in the form of the large steel pipelines that were useful only for shipping gas, and Europe was the only viable export market at the time. Mark H. Hayes and David G. Victor, 'Politics, Markets, and the Shift to Gas: Insights from the Seven Historical Case Studies', in Victor, Jaffe and Hayes, eds., n. 18, p. 345.

50 'The Gas Exporting Countries Forum'. http://www.gecf.org/Resource/GECF-History-File.pdf. Accessed on 24 October 2014.

51 Gawdat Bahgat, 'Prospects for a Gas OPEC', *Middle East Economic Survey*, Vol. 53, no. 2, 12 January 2009.

52 Vijay Vaitheeswaran, 'Oil: Think Again', *Foreign Policy*, November–December 2007, p. 26.

53 'Medium Term Oil and Gas Markets 2011', *International Energy Agency*, pp. 255–257. Also see Derek Brower and Kirk Sowell, 'Pipelines, Not LNG, for East Med Gas', *Petroleum Economist*, Vol. 81, no. 6, July–August 2014, pp. 26–27.

54 Amy M. Jaffe and Ronald Soligo, 'Market Structure in the New Gas Economy: Is Cartelisation Possible?', in Victor, Jaffe and Hayes, eds., n. 18, p. 458.

55 Putin's oft-repeated response to the idea has been,

We have no plans to create some kind of cartel, but I think it would be a good idea to coordinate our activities, especially in the contract of achieving our main aim of ensuring certain and reliable supply of energy resources for our main consumers. Quoted in Jonathan Stern, 'Gas OPEC: A Distraction from Important Issues of Russian Gas Supply to Europe', *Oxford Energy Comment*, February 2007. http://www.oxfordenergy.org/wpcms/wp-content/uploads/2011/01/Feb2007-GasOPEC-JonathanStern.pdf. Accessed on 13 June 2013.

56 Henry Hub, located in Louisiana in the US, is a gas-distributing and price-setting point. Unlike the traditional long-term bilateral gas deals, Henry Hub prices are market dictated.

57 James Crabtree, 'Asia Gas Buyers Club Threatens Canada's Energy Plans', *Financial Times Blogs*, 15 January 2014. http://blogs.ft.com/beyond-brics/2014/01/15/asia-wide-gas-buyers-club-throws-spanner-in-cana das-energy-plans. Accessed on 31 January 2014.

58 Gal Luft, 'Building an Asian Energy Buyers Club', *Wall Street Journal*, 2 September 2013.

59 'India, Asian Countries to Launch Gas Buyers' Club', *Business Standard*, 3 September 2013. In a strange coincident, Tripathy's announcement came almost parallel with the meeting of the Asia-Pacific energy ministers in Beijing that was expected to discuss an energy buyer's club.

60 Nidhi Verma and Jo Winterbottom, 'Asian LNG Buyers Come Together in Bid to Secure Lower Prices', *Financial Post*, 3 December 2013.

61 Anilesh A. Mahajan, 'A Super Cool Idea', *Business Today*, 5 January 2014.

62 'GAIL, Japan's Chubu form Gas Buyers Club', *Times of India* (New Delhi), 30 January 2014.

63 'Japan is not that far away. So if you can't even get Japanese buyers to a major conference, what chance does the Asian buyers' club see in this regard?' asked Noel Tomnay, head analyst of global gas research at Wood Mackenzie. Meeyoung Cho and Jane Chung, 'Asian Gas Buyers Trying to Break Out of Rigid Market Structure', *Reuters*, 4 April 2014.

64 David Capole, 'Global Natural Gas Explosion: Clean, Cheap Product/Costly, Destructive Process', *Economy Watch*, 5 April 2011. http://www.economywatch.com/economy-business-and-finance-news/global-natural-gas-explosion-clean-cheap-product-costly-destructive-process.05–04.html. Accessed on 13 June 2013.

65 The author does not explain how the feat was accomplished in the early nineteenth century with the technology and tools available then. Michael Levy, 'Splitting Rock vs. Splitting Atoms: What Shale Gas Means for Nuclear Power', *Bulletin of Atomic Scientists*, Vol. 68, no. 4, July–August 2012.

66 Ibid.

67 Edward L. Morse, 'Welcome to the Revolution: Why Shale Is the Next Shale', *Foreign Affairs*, May–June 2014. http://www.foreignaffairs.com/articles/141202/edward-l-morse/welcome-to-the-revolution. Accessed on 14 October 2014.

68 'Shale Oil and Gas Resources Are Globally Abundant', *Energy Information Administration*, 2 January 2014. http://www.eia.gov/todayinenergy/detail.cfm?id=14431

69 Levy, n. 65.

70 Iana Dreyer and Gerald Stang, 'The Shale Gas "Revolution": Challenges and Implications for the EU', *ISS Alert*, February 2013. http://www.iss.europa.eu/uploads/media/Brief_11.pdf. Accessed on 2 October 2014.

71 Alan Riley, 'The Shale Revolution's Shifting Geopolitics', *The Hindu*, 31 December 2013.

72 Christopher Ross, n. 33.

73 France, Bulgaria and the Czech Republic have banned fracking. Germany is not issuing production permits, although debate and research continue. The United Kingdom, a declining conventional gas producer, has authorised fracking, albeit with strong regulations. In late 2012, the EU Parliament, while not calling for a ban on fracking, did call for a robust regulatory framework. The EU has set itself the target to reduce its carbon emission by 20 per cent from where it stood in 1990. Dreyer and Stang, n. 70.

74 Igor Alexeev, 'Fracking Fantasies: Has the Shale Bubble Already Burst?', *Economy Watch*, 26 August 2013. http://www.economywatch.com/economy-business-and-finance-news/fracking-fantasies-has-the-shale-bubble-already-burst.26–08.html. Accessed on 26 September 2013.

75 US environmental authorities have documented a number of such accidents, in which wastewater has harmed the environment, even though the gas industry's engineers insist that contact with groundwater is highly unlikely, because the layers of rock containing the gas are so much deeper. Capole, n. 64.

76 Bill McKibben, 'Actions Speak Louder Than Words', *Bulletin of the Atomic Scientists*, Vol. 68, no. 2, March–April 2012, pp. 1–8.

77 Alexander Orlov, 'Global Energy: New Geopolitical Equations', *International Affairs* (Moscow), Vol. 58, no. 6, 2012, p 190.

78 For example, experts have long talked about the possibility of capturing carbon dioxide emissions from coal and burying them underground, an approach known as carbon capture and sequestration, or CCS. The same approach can be applied to natural gas. Levy, n. 65.

79 Most readily among the unemployed strata of society.

80 Stimulating industrial and manufacturing growth and reducing the proportion of gross domestic production (GDP) spent on energy imports are the immediate positive fallouts of the shale. David Hastings Dunn and Mark J. L. McClelland, 'Shale Gas and the Revival of American Power: Debunking Decline?', *International Affairs* (London), Vol. 89, no. 6, 2013, pp. 1411–1428.

81 Dreyer and Stang, n. 70.

82 Orlov, n. 77.

2

GAS PIPELINE
Commodity, container and carrier

> In Saudi Arabia one finds a combination of terrains; some pipelines are laid between the hundreds of huge sand dunes of the Rub' al-Khali (The Empty Quarter), other pipelines buried in the wet and salty soil of Sabkhat Jayb Uwayyid (a salt flat), some pipelines cross underneath the hard rocky mountains of Al-Sarawat – that stretch across the entire western region of the country – and others are laid on top of the seabed of the Arabian Gulf.[1]
>
> More than 95 percent of gas produced and three quarters of gas traded is distributed via pipelines directly from supplier to consumer, and gas-to-liquids technology is unlikely to change these ratios substantially by 2020.[2]

These quotes describe the sheer physical work, immense financial commitment and the most complex technological input into the task of laying the pipelines, and yet hold the endeavour to be unavoidable for most of the natural gas distribution at least till 2020. This chapter will study the construction and utility of the gas pipelines.

Bamboo to the barrel: gas through the millennia

The Chinese were the first to have used gas. As early as 940 BC, people in China piped natural gas through hollow bamboo poles to the seashore, where they used it to boil ocean water and collect salt. Some experts say that the Chinese drilled gas wells as deep as 600 metres. Japanese wells were reported around 600 BC.

Other ancient civilisations noticed the escape of natural gas from the ground and discovered that it would burn. Temples were built to house these mysterious 'eternal fires', regarded by visitors with a mixture of reverence and superstition. Later reports note pillars of fire and bubbling magic water that would burn like oil. The concept of eternal fire is global and has religious as well as emotional connotations. The Zoroastrian fire temples and the eternal flame of Olympic are the most noted examples of fire as symbols of purity and continuity through ages.

The emerging gas industry in the US and Europe relied on heating coal to manufacture gas. Early distribution systems used wooden pipes, which were replaced in time by metal pipes or 'barrels' (made in the same way as the navy's gun barrels). By 1819, London alone had nearly 482.7 kilometres (km) of gas pipes supplying more than 50,000 burners.[3]

Gas pipelines are large-diameter pipes with welded joints. As gas is transported at high pressure over long distances, compression stations are built at strategic points along the pipeline to maintain the pressure. Gas engines and turbines are used to power the compressors. In addition, safety shut-in valves are included at critical places in the pipeline, particularly near high-consequence areas.[4] The passage of gas is contingent upon computer modelling that enables the pipeline to extract gas that is needed and not leave extra storage built up in the pipeline.

The pipeline cycle runs along these steps: the small flowlines carry gas from the wells to gas-treating and gas-processing facilities. Here, water content and sulphur are separated by computer and sophisticated communication techniques. The steel pipelines are either seamless or longitudinally welded. The spiral pipes are diameter adjustable. The internal coating of the pipes reduces the internal pipe's roughness. The external coating prevents corrosion.

The weight coating or anchoring is done to the sub-sea pipes to overcome the force of buoyancy and prevent the pipe from floating to surface. Care needs to be taken to bury the pipes at depths to avoid ship anchors, fishing gear and such other obstructions. At times, there are decommissioned offshore platforms that need to be avoided. The pipes need to be secured against natural disasters like earthquakes.[5] Greater water depths and larger pipes can lead to buckling. Permafrost conditions make it necessary to build some sections of the pipe above ground. Generally, sub-sea lines at depths of over 3,000 km are regarded as uneconomical and risky as the pipeline and compression installation and maintenance are expensive and difficult.

Laying the pipelines on the mountainous regions is no less difficult. The rocky terrain, dangerously steep slopes, potential landslide risks, ravines, flash rains in winter, gravity and its effect on the material being moved as well as the equipment moving it are serious considerations in the mountains as compared to the normal, open and flat construction spreads. The most serious challenge is keeping the excavators, workers and trucks from rolling over on steep mountain sides.[6] Landslides and seismic areas are additional hazards.

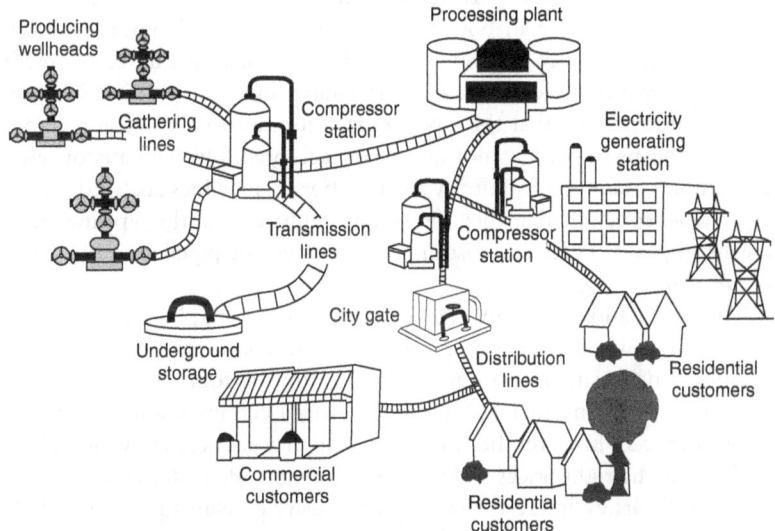

Figure 2.1 Natural gas delivery network.

Source: Based on Canadian Energy Pipeline Association. http://www.cepa.com/about-pipelines/types-of-pipelines/natural-gas-pipelines

The pipes need to be secured from a leak that can range from a pinhole leak to a catastrophic failure. The sour gas[7] pipelines are particularly vulnerable to a leak. The prevention, mitigation and control of a leak begin with the process of selection of materials that meet project-specific requirements, intended design life, costs and environmental considerations.[8] Once the pipes are operational, the process of 'pigging'[9] is done. The pigs are inspection gauges that can perform various maintenance operations – from inspection to cleaning – without stopping the pipeline flow. The current generation of 'smart pigs' can detect corrosion in the pipeline and is thus relied on for leak detection. For larger pipes, the industry relies on a more advanced infrastructure monitoring system, where remote hubs relay data back to central monitoring point, using fibre-optic cable or other communications equipment. RealSens is an advanced remote sensing pipeline detection technology that detects leaks over an entire pipeline network.[10]

Pipeline economics and advantages

The gas pipeline, like any cross-border trade, benefits all parties involved: it leads to reduction in energy inequality in the region; it is a win–win scenario for importing, exporting and transit countries; it necessitates infra-structure development in border areas, which are usually remote, in difficult terrain and away from important cities and mainstream economy. It stimulates economic activities, and it facilitates improved political relationship.[11] The assertion that the pipelines lead to improved political relations or are a means to peace is not universally accepted. On the basis of seven case studies examined and nearly all cross-border pipelines analysed, Hayes and Victor conclude that they showed no evidence that the pipelines are a means to peace. Peace and institutions preceded the pipelines, rather than the reverse.[12]

The International Energy Agency (IEA) takes a middle ground in this debate. It maintains that the pipelines tie the consumer to the supplier, creating a negotiating position which sometimes favours the supplier and sometimes the consumer but always involves a certain amount of trust.[13] Customers can, however, be reasonably sure that gas keeps flowing as long as they pay the right prices and the gas resource is adequate, since it is generally in all parties' interests to keep an expensive pipeline fully utilised.[14] As noted in the previous chapter, the relations may not be as smooth over the entire operation; once the pipelines become operative, the supplier country has an upper hand as the exporting company has already put in most of the capital required and has few options left to get out of the deal.

The pipelines are costly to build because of expenses for the high-strength, large-diameter pipes; large compressors; and land acquisition. Because of the huge investments committed, pipeline companies strive to operate as close to the maximum capacity as possible. And also because if the pipelines have to be shut down, the producing and receiving facilities and refineries also have to be shut down, as gas cannot be easily stored. The economics of pipelines explains why most are built to an initial capacity level, with the option to expand capacity at a later date if circumstances warrant. This staged construction avoids excess capacity, dragging down profitability. Others are designed for throughput to increase in stages, should demand warrant, where stages call for more pump stations, looping the line or introduction of chemical reagents known as drag reduction agents (DRAs).

There are operational costs, since natural gas is far more costly to move over long distances than is oil. According to one estimate, while it costs roughly $0.50 to move a barrel of oil 1,000 km by pipeline (depending on the diameter and number of pumping stations), it would cost between $2.50 and $4.00 to move the same amount of energy in the form of natural gas over a similar distance.[15] Economy of scale demands that the line must be large; at the same time, cost-effectiveness can be achieved only if the capacity of the line is fully utilised within a few years of completion. Gas, as a result of the storage difficulties, needs to be transported immediately to its destination after production from a reservoir.

The viability of a pipeline is inextricably linked to the costs of producing gas. If the production costs are low, then the profits would be high. Conversely, if the gas is high cost at the input point, the profit margins would be narrow. In addition, a pipeline must be filled quickly to remain economically viable. An empty or half empty pipe is uneconomical. There must be a matching production potential and market demand. Asymmetrical demand and supply lead to avoidable economic costs. Another vital question is whether there are smaller-scale, incremental options that compete against the pipelines. The spectre of 'incrementalism' is real. A pipeline can potentially be undercut, if not superseded entirely, by alternative pipeline routes. Since the competitors are generally cheaper and quicker to build, the original pipeline suffers the spectre of death by 'a thousand cuts'.[16]

Pipeline passages

The transnational gas pipelines involve two sovereign independent states and very often more than two. The trans-border lines are extending to be transcontinental lines with advances in technology. The longer the length of the pipelines and greater the number of countries through which pipelines pass, the stronger the possibility of short- and long-term interruptions in the flow of oil or gas, and therefore the higher the vulnerability of these pipelines.[17] The smooth functioning of a line, in this sense, requires much more than a functioning infrastructure. The human dimension of the success of a line is equally important. In addition to a committed investor, domestic politics in the countries involved; bilateral relations between the supplier and producer countries; relations among the supplier, producer and the transit countries – all these must line up. And even then, 'caveat emptor' is the best strategy.[18]

The transit states are indispensable if the line has to cross its territory to reach the importing state. Their demand for rent for the lease of the territory for the passage of the line is justified. The amount of rent demanded is where the negotiations begin. The most important ingredient in the final settlement reflects the presence or absence of an alternative, be it an alternative route or be it the option of LNG. No transit country is purely a transit facilitator. Most of them are gas consumers as well. In these cases, the amount of gas that they are allowed to keep for their use is an inevitable clause in the agreement.

Till recently, Ukraine was the transit state for most of the Russian gas to Europe. Ukraine was the choking point as well as leaking point of the gas in transit. Russia is a transit country for the Turkmen gas – at the other end of the chain. All the Turkmen gas exports outside China and Central Asia pass through Russia, putting the latter in complete control of Europe-bound exports. The triangular chain has gone though many inevitable hiccups. The following chapters will take up the issue in greater detail.

In a case where the transit state is also a gas producer, the situation is even tougher to resolve. Its primary interest obviously is to sell its own gas rather than lease its territory for another's gas exports. It could refuse permission to let the line pass through its territory, create obstructions in the construction of a line or resort to reduce the supply to secure a share of the market for itself. The Saudi objection to the Qatari gas line to Oman is a case in point.

With the gas line in place and running, the producers and the importers are constantly required to be alert to interruptions. The transit state retains its privileged position of the passage throughout the life of the line. Its demands may go up, its relations with any of the parties may develop frictions and its own domestic situation may undergo a change – either due to a regime change, a secessionist movement or a failing/failed polity. The project companies routinely undertake development projects in the pipeline neighbourhoods like providing employment, electricity, water, health services, schools, roads and parks. Funding politicians and the non-governmental organisations is not uncommon. In addition, transit states are kept interested in the project by plans of infrastructure expansion and increase in throughput that would lead to higher rents to the transit states.

Manipulating and partnering with influential actors in the transit state is often resorted to not just in personalised polities but in highly institutionalised developed states as well. The most glaring case is the former German chancellor Gerhard Schroeder, who on demitting office, became the chairman of the supervisory committee of the Russian North European Gas Pipeline Company, with an overall responsibility for building the pipeline under the Baltic Sea.[19] As he had promoted the Nord Stream pipeline from Russia to Europe, when he was the chancellor, his appointment was seen as a payback and roundly criticised.

Pipelines versus the LNG

There are a number of options for transporting gas, which include pipelines; liquefied natural gas (LNG); compressed natural gas (CNG); gas to solids (GTS), that is hydrates; gas to power (GTP), that is electricity; and gas to liquids (GTL), with a wide range of possible products, including clean fuels, plastic precursors or methanol and gas to commodity (GTC), such as aluminium, glass, cement or iron.[20] Out of these, pipelines and the

LNG are the most widely used options. LNG involves cooling the gas to minus 160 degree Celsius that occupies 1/600 of its gaseous form. The liquid gas is transported in specially designed cryogenic carriers, which have double hulls to protect the gas from leakage, accidents or damage. Once delivered to the destination, it is warmed back to its original gaseous form and distributed by normal gas pipelines. The exporting liquefaction terminals and the importing re-gasification terminals are specially equipped to handle the cargo.

Whereas the pipes have transported gas through the millennia, the LNG is comparatively a recent mode of gas transport. LNG is estimated to be economical if gas reserves at the well head are at least 3 to 5 tcm in size and the pipelines to the market are longer than 1,000 kilometre. Today, 95 per cent of produced gas and 75 per cent of traded gas is transported via pipelines. GTL technology is unlikely to change these ratios substantially by 2020. According to an energy analyst:

Pipelines will be the most economical way to transport oil, gas and petroleum products overland in the foreseeable future. Most future advances in oil and gas pipeline technology will also be gradual improvements in efficiency and capability. Few dramatic changes are expected. Most of the technology required in the coming years will not be revolutionary, but the need for innovative approaches to lower cost and higher efficiency has never been greater, he adds.[21]

What matters for the LNG is the availability of a large gas resource near a coastline and the proximity to any market – close enough to make shipping feasible and yet far enough to make LNG the preferred option over a pipeline.[22] Such privileged sites are becoming rare, with production now moving to less accessible sites and landlocked areas. With this, pipelines remain a preferred option. The pipes are much more energy efficient as they use only 1 per cent of the fuel they carry. The tankers, trucks or trains, in comparison, are high in fuel consumption. And lastly, the LNG ships and terminals are considered more vulnerable to terrorism[23] and, therefore, more unsafe than the pipes. An LNG tanker exploding in a seaport could be as devastating as the 9/11 attack in the US. A very large container carrier (VLCC) or an LNG tanker could be rammed into a major port or a large bomb could be set off by a suicide bomber or via a remote control on board a large cruise ship.[24] Though their enormous size makes them hard to sink, it also makes them easy targets for swifter, smaller boats. A tanker's maximum speed is about 14 knots an hour, while a small boat can reach 70 knots.[25]

There are some points on which the piped gas loses out to the LNG. Gas has sheer volume with thin energy content. Distance is a critically limiting factor for a gas pipeline. Even if all other considerations are favourable, it is much more costly to move energy in the form of natural gas than the comparable amount in the form of liquid. Liquid gas reduces the volume and concentrates the energy content. Transporting the liquid is transporting less of a cargo. Another major advantage of the LNG is that it avoids passing through state territories. It is not routed through the transit countries. Although the LNG tankers ply the territorial waters of transit states, they have a variety of routes available should a transit country choose to hold up a trading project. Avoiding land passage does not, however, automatically translate into safe passage. The LNG has to pass through much more hazardous choke points in the world in its journey to the market. The Strait of Hormuz between Iran and Oman is the only sea passage from the Gulf to the open sea. The Strait of Malacca between Malaysia and Indonesia connecting the Pacific Ocean to the Indian Ocean is one of the world's most strategically important choke points. The gas supplies from the Gulf to Asian markets like China, Japan and South Korea pass through this strait.

In recent years, pipeline economics has benefitted from new technology. The principal improvements have been in the availability of improved steels that make high-pressure operations feasible and in marine pipe-laying techniques that have made pipelining somewhat more competitive with LNG than it might have been 10 to 15 years ago. However, the recent inflationary pressures on material costs appear to be more severe for pipelines than for LNG.[26] An alternative view considers the LNG expenses to be still large in comparison; in fact, it considers the LNG facility to be a floating pipeline that is generally not built without locking in supply or a market.[27]

The recent technological advances have also benefitted the LNG in the form of floating production, storage and offloading (FPSO).[28] It is another type of LNG tanker with gas liquefaction facilities equipped on the top of the tanker. This is the most advanced technology currently being tested in the LNG industry. It is also the most suitable for stranded gas fields because this technology opens the door for projects deemed uneconomical due to the remoteness from land or insufficient reserves (generally less than 2 tcf or 0.057 tcm proven reserves). Although there is no FPSO currently under operation, several companies are keen to develop this technology to materialise their LNG projects – the US Shell and the Australian Woodside, among them.

The CNG that contains volumetric density amounting to 42 per cent that of the LNG is yet another transportation option. A giant airship flying low and slow over under-populated terrains or seas and avoiding risks of fire is one more such option. Both of these are futuristic ideas, whose time has not yet come.

Security/insecurity of the pipelines

The energy exploration and extraction is a hazardous activity at every stage of its operation from the well head to the market. As the energy demand and economic rewards go up, the scramble to get to the resource faster becomes more intense. Deeper underground, deeper ocean floor, distant offshore, higher mountains, harder rocks, remoter and more inaccessible areas – all these come progressively within the explorers' reach.

The oil/LNG tankers represent the weakest link in the energy chain, followed by the vulnerable choke points like the narrow straits and waterways close to the coasts that the seaborne commerce has to traverse. The oil and gas pipelines are the next points of vulnerability in the energy trade. There are more gas and oil pipelines crossing the international border than ever before. Many of them are mega projects envisioning lines that run across several countries – at times, connecting an entire continent, at others, going beyond across water bodies. Very often, they are submerged below water and not always hugging the coastline.

While it is easy to draw lines on a map, building pipelines in adverse permafrost conditions, undersea, and through deserts and forests is a complicated and expensive business.[29] Natural disasters and human interventions add to the endeavour. Extreme weather events like typhoons, tropical cyclones and seismic activities, for instance, lead to costly and deadly disasters. An increased likelihood of unauthorised activities like fishing, diving or tourism in close proximity of the pipeline installations presents safety risks. The numbers of decommissioned offshore oil and gas platforms increase as existing, older platforms near their end of life. The incidents of derelict facilities will significantly increase into the future due to the large number of new installations. The risks posed by abandoned rigs include hazards to navigation and other users of the area. The internationally mandated 55-metre safety zones are not wide enough to provide adequate space to warn or intercept intruders. In crowded waters the results of incidents can quickly extend beyond individual national jurisdictions.[30]

Even though the pipelines are mostly buried underground, they are laid just below the surface and their route is well marked to facilitate maintenance, making them prone to disruption. In this situation, the entire length of the pipeline would need to be fenced off on both sides to deny easy access to prospective saboteurs. Since the wire fencing can be easily cut, it would need to be kept under electro-optical surveillance throughout its length, combined with continuous physical patrolling. All these measures would cost a massive amount to implement and would still not guarantee 100 per cent security.[31] Once attacked, huge expenses would have to be incurred and technical skills would have to be deployed.

In the circumstances, the gas pipelines have assumed a critical role in international security affairs. American soldiers are now helping to defend such conduits against attacks in Iraq, Colombia and Georgia. President George W. Bush had authorised the deployment of US military personnel in Georgia to help train the Georgian troops that would be responsible for protecting the Baku–Tbilisi–Ceyhan (BTC) and the Baku–Tbilisi–Erzurum (BTE) lines. Pipeline protection has become a major concern for Saudi Arabia, Sudan, Algeria, Nigeria, Burma and other strife-torn producers. Located roughly 80 km north of Doha, Ras Laffan in Qatar is a heavily guarded industrial city producing liquefied gas and gas to liquids. Entry to the city is severely restricted and photography is forbidden. The Qatar government takes all possible precautions against sabotage.[32] Against the background of turmoil during the Arab Spring in 2011, Saudi Arabia began to deploy a 35,000-strong specialised protection force to guard crude-processing plants, oil fields and pipelines.[33] The BTC and the BTE pass too close to volatile breakaway regions in both Georgia and Azerbaijan, making them vulnerable to sabotage that could cause a catastrophic spill. The Georgian government, in the situation, has dropped original plans to patrol the BTC/BTE route with unmanned reconnaissance aircraft in favour of small, roving anti-terrorist squads.[34]

The Indian parliament has passed the Petroleum and Minerals Pipelines (Acquisition of Right of User in Land) Amendment Bill 2010 that provides for a minimum of 10 years' rigorous imprisonment for acts of terrorism and makes such activities punishable by even death sentence.[35] The Gas Authority of India Limited (GAIL) has decided to deploy drones to guard pipelines to raise safety standards after a major explosion in a gas pipeline.[36]

The large fixed energy installations are difficult targets for terrorist attacks, although they do happen. Stolen gas has no resale value, and the gas pipelines are not attacked for theft, as are the oil pipelines. The gas pipelines are attractive targets for the violent dissident movements at home. A disruption of the gas supply strikes a direct blow against the regime and its foreign backers. It cripples the economy. It leads to power outages inciting opposition to the government's incompetence. In addition, gas from a ruptured line ignites, creating spectacular fireballs and an impact many multiples of the ammunition used.[37] The demonstrative value of such an attack is great. Since the pipes carrying gas have pressure in them to move the gas along, any attack would invariably lead to an explosion. It could either be a suicidal vandal who could attempt it or be done with the collusion of the operators to turn off the gas before the pipe is blown up.[38]

The Iraqi militants resorted to such acts against the occupying forces after 2003 US war in the country. The violent Pakistani opposition and also the opposition in Yemen[39] have regularly undertaken such acts. In

Colombia, a pipeline running from the north to the eastern Caribbean port of Covenas has been ruptured so many times that it has been nicknamed 'the flute'.[40] In Nigeria, over 200 incidences of crude oil and gas pipeline vandalism were recorded in six months till February 2015, according to the Nigerian federal government.[41] Until the Arab Spring, Israel was a buyer of Egyptian gas through a spur connected to the Arab gas pipeline via the Sinai. The pipeline was deeply unpopular in Egypt. After the overthrow of the Hosni Mubarak government, the infrastructure was repeatedly attacked, cutting the flows both to Israel and to Jordan, another recipient of the Egyptian gas.[42] In January 2013, militants raided Algeria's Amenas gas field, sparking a crisis that ended with the deaths of at least 37 hostages. The Islamic State of Iraq and Syria (ISIS) is in control of gas fields in central Syria, although the lack of pipelines for transporting the gas has left the gas stranded. It is reported to have blown up a gas pipeline from eastern Syria to the suburbs of Damascus that generated electricity and provided heating in homes.[43] The instances quoted here do not include all the acts in all parts of the world that have happened. They are only examples of the trend and expanse of the phenomenon.

What is extremely alarming is the cyber-attacks on the computer networks of gas pipeline systems in the US.[44] The US Congress passed the Cyber Security Act in February 2012 specifying threats of interruption to the life-sustaining services, including energy, water, transportation, emergency services or food. It asked for annual reports from the Department of Homeland Security summarising major cyber incidents and aggregate statistics on 'the number of breaches of networks of Executive agencies, the volume of data exfiltrated, and the estimated cost of remedying the breaches'.[45]

Beginning 29 March, that is barely a month after the Congressional Act, at least three confidential 'Amber' alerts – second-most sensitive next to the 'Red' alerts – were issued by the Department of Homeland Security. The alerts warned of a 'gas pipeline sector cyber intrusion campaign' against multiple pipeline companies. The wave of cyber-attacks had apparently begun four months earlier already and was continuing till May, when they were confirmed by the US officially. The Industrial Control Systems Cyber Emergency Response Team (ICS-CERT), an arm of the Department of Homeland Security, reaffirmed the fact in an 'incidence response' report.[46]

Since the gas pipes, nuclear power plants, water systems and subway systems are all connected to the computer systems, there could be devastating consequences if they are hacked.[47] The security of pipelines came in for special consideration in the cyber security programme. In November 2014, the Department of Homeland Security formed the Pipeline Working Group and the Pipeline Sector Coordinating Council (PLWG/

PSCC) with a mandate to improve safety, security and resiliency in the pipeline systems 'designed to have minimal disruptions and the recovery of operations in the event of an incident or disaster'.[48] The extra vigilance accorded to the pipelines demonstrates the priority of the pipeline infrastructure and has paid off as there has been no reported sabotage till now.

According to a Joint UNDP/World Bank Energy Sector Management Assistance Programme (ESMAP), the destruction and the disruption of pipelines in a few cases have cast a much greater shadow than their actual numbers justify. Often, the negative perceptions have inhibited the operation of existing lines and the building of new ones. The perceived risks increase the cost of finance and seriously impact the delivered cost of the fuel.[49] The ESMAP report prescribes interim state support for pipeline projects and cautions against the tempting argument that the state should set the context and then move aside to allow the fullest involvement of the private sector.

Lastly, there are state-level political threats to the perfectly secure and functioning pipelines. Sanctions on exporting countries like Russia and Iran may render a pipeline worthless. Civil wars like in Syria may disrupt the flow. Domestic demands may use up the supply meant for exports like in Russia. Secessionist/dissident movements like the Aceh Rebellion in Indonesia in the 1950s, the Chechen rebels in Russia in the 1990s and the Kurdish region in Turkey may adversely impact the lines or necessitate change in their routes to avoid trouble. The failing/fragile state with no control over much of its territory may not be able to provide necessary security. Wars, as in Iraq, may make the lines targets of external attacks. Tribal rivalries as in Yemen may make the lines pawns in the ongoing disputes. The insurgents/separatists may resort to kidnappings and extortion, demanding protection money to let the pipe pass without sabotage. The regional dynamics also come into play, like when the Saudis objected to the Qatari gas reaching the United Arab Emirates and Oman. And the perennial hagglings over the price between a buyer and a seller could lead to temporary shutdowns of supply.

Recapitulation and leads

The task of laying the gas pipelines involves sheer physical work, immense financial commitment and the most complex technological input. The Chinese were the first to have piped gas in hollow bamboo poles as early as 940 BC. Today, 95 per cent of produced gas and 75 per cent of traded gas is transported via transnational pipelines. Gas pipelines are large-diameter pipes with welded joints. As gas is transported at high pressure over long distances, compression stations are built at strategic points along the pipeline to maintain the pressure. Gas engines and turbines are used to power the compressors. Like any cross-border trade, the pipelines benefit the exporting, transit and importing countries. They necessitate infrastructure development and generate employment along the route. At the same time, they entail hard bargains, disputes and conflicts, resulting in interruptions in the flow of gas. The longer the length of the pipelines and greater the number of countries through which they pass, the stronger the possibility of interruptions and therefore the higher the vulnerability of these pipelines. A committed investor; domestic politics in the countries involved; and the relations among the exporting, transiting and importing countries have a bearing on the successful operation of the pipelines. Pipelines and the LNG are the most widely used options for transporting gas. LNG involves cooling the gas to minus 160 degree Celsius that occupies 1/600 of its gaseous form. The liquid gas is transported in specially designed cryogenic carriers, which have double hulls to protect the gas from leakage, accidents or damage. Once delivered to the destination, it is warmed back to its original gaseous form and distributed by normal gas pipelines. The exporting liquefaction terminals and the importing re-gasification terminals are specially equipped to handle the cargo. LNG is estimated to be economical if gas reserves at the well head are at least 3 to 5 tcm in size and the pipelines to the market are longer than 1,000 km. Natural disasters like cyclones, seismic activities and human interventions like fishing and diving present safety risks to the pipelines. They are also vulnerable to terrorist attacks including cyber-attacks on the pipeline's computer networks.

Notes

1 Husam H. Al-Hasan, Faisal A. Qari and Syed M. Badruddoza, 'Saudi Aramco Faces the Desert Pipeline Challenge', *Pipelines International*, September 2013. http://pipelinesinternational.com/news/saudi_aramco_faces_the_desert_pipeline_challenge/083147/. Accessed on 2 June 2014.
2 'Rising Powers: The Changing Geopolitical Landscape', *Report of the National Intelligence Council's 2020 Report*. http://www.globalsecurity.org/intell/library/reports/2005/nic_globaltrends2020_s2.htm. Accessed on 20 September 2013.
3 Rebecca L. Busby, ed., *Natural Gas in Non-Technical Language* (PennWell, Tulsa, 1999), pp. 5–6.
4 Kelvin T. Erickson, Ann Miller, E. Keith Stanek, C. H. Wu and Shari Dunn-Norman, 'Pipelines as Communication Network Links', *Science.gov*, 14 March 2005. http://www.osti.gov/scitech/servlets/purl/839987. Accessed on 29 October 2014.
5 In April 2014, The US Federal Emergency Management Agency (FEMA) concluded in a report that natural gas pipelines located beneath the ground throughout the US were not designed to withstand the seismic force of

a high-powered earthquake. It is said to be especially true for those areas of the US where it may be unthinkable that an earthquake could possibly happen, specifically in the central and eastern portions of the country. Quoted in Stephanie Tapley, 'Natural Gas Pipelines May Impose More Harm Than Good', *Guardian Liberty Voice*, 9 April 2014. http://guardianlv.com/2014/04/natural-gas-pipelines-may-impose-more-harm-than-good/. Accessed on 30 June 2015.

6 Paul Trundle, Pepspen Head of Pipeline Engineering. Quoted in Stephanie Chan, 'Ain't No Mountain High Enough to Stop Pipeline Construction', *Pipeline International*, March 2012. http://pipelinesinternational.com/news/aint_no_mountain_high_enough_to_stop_pipeline_construction1/067035. Accessed on 13 May 2013.

7 Sour gas contains a significant amount of hydrogen sulphide. It can be corrosive, flammable and fatal to humans and animals. The gas that is free of this substance is called sweet gas.

8 David Newman, 'Sour Gas Pipelines – How Do We Deal with Them?', *Pipeline International*, December 2011. http://pipelinesinternational.com/news/sour_gas_pipelines_how_do_we_deal_with_them/65073. Accessed on 30 September 2012.

9 The device got its name from the squealing noise it emitted while travelling through the pipeline.

10 'Pigs in the Pipeline: Why We Need Them, Where They Fail', *Economy Watch*, 12 August 2013. http://www.economywatch.com/economy-business-and-finance-news/pigs-in-the-pipeline.12–08.html. Accessed on 13 September 2013.

11 Nathan Hippu Salk Kristle, Sanket Sudhir Kulkarni and Dilip R. Ahuja, 'Pipeline Politics – A Study of India's Proposed Cross Border Gas Projects', *Energy Policy* (Elsevier, Philadelphia), Vol. 62, 2013, p. 148.

12 Mark H. Hayes and David G. Victor, 'Politics, Markets, and the Shift to Gas: Insights from the Seven Historical Case Studies', in David G. Victor, Amy M. Jaffe and Mark H. Hayes, eds., *Natural Gas and Geopolitics: From 1970 to 2040* (Cambridge University Press, Cambridge, 2006), pp. 346–347 ff.

13 "Mutual interdependence" is a more appropriate term in the context than the "mutual trust" used here. Chandrashekhar Dasgupta, 'Ukraine, India and Energy Security', *Gas and Oil*, 18 January 2006. http://www.gasandoil.com/news/2006/02/nts60608. Accessed on 31 November 2013.

14 'Natural Gas Market Review 2006: Towards a Global Gas Market', *International Energy Agency*, 2006, p. 20.

15 Maureen S. Crandall, *Energy, Economics and Politics in the Caspian Region: Dreams and Realities* (Praeger Security International, Westport, 2006), pp. 24 and 175 ff.

16 Thomas R. Stauffer, 'Caspian Fantasy: The Economics of Political Pipelines', *Brown Journal of World Affairs*, Vol. VII, no. 2, Summer–Fall 2000, pp. 63–78.
   The original pipeline, in this situation, would be deprived of its economy of scale, if not a share of the business.

17 Hooman Peimani, *The Caspian Pipeline Dilemma: Political Games and Economic Losses* (Praeger, Westport, 2001), p. 45.

18 A Latin phrase that says 'Let the Buyer Beware'.

19 Schroeder was denounced for this act across the German political spectrum. Craig Whitlock and Peter Finn, 'Schroeder Accepts the Russian Pipeline Job', *Washington Post*, 10 December 2005.

20 Saeid Mokhatab, William A. Poe and James G. Speight, *Handbook of Natural Gas Transmission and Processing* (Elsevier, Burlington, 2006), p. 19. http://www.academia.edu/3126636/HANDBOOK_OF_NATU RAL_GAS_TRANSMISSION_AND_PROCESSING. Accessed on 29 October 2014.

21 John L. Kennedy, *Oil and Gas Pipeline Fundamentals* (PennWell Publishing Company, Tulsa, Oklahoma, 1993), 2nd Edition, p. 357.

22 Mark H. Hayes and David G. Victor, 'Politics, Markets, and the Shift to Gas: Insights from the Seven Historical Case Studies', in Victor, Jaffe and Hayes, eds., n. 12, p. 342.

23 It could be one of the reasons that the energy companies in the US had constructed only four terminals capable of handling the nation's needs till the early years of this century. Dilip Hiro, *Blood of the Earth: The Battle for the World's Vanishing Oil Resources* (Penguin Books, New Delhi, 2008), p. 252.

24 Vijay Sakhuja, 'Energy Transportation Security in the Bay of Bengal', in Sudhir Devre, ed., *A New Energy Frontier: The Bay of Bengal Region* (Institute of Southeast Asian Studies, Singapore, 2008), p. 149.

25 That is what happened on 8 October 2002, when Limburg, a French oil tanker, was hit off the coast of Yemen. As oil kept gushing into the Arabian Sea from the 56,000-tonne carrier, oil prices rose by 24–26 cents per barrel.

26 James T. Jensen, 'The Future of Gas Transportation in the Middle East Region: LNG, GTL and Pipelines', in *The Gulf Oil and Gas Sector: Potential and Constraints* (Emirates Center for Strategic Studies and Research, Abu Dhabi, 2006), p. 267.

27 Brenda Shaffer, 'Natural Gas Supply Stability and Foreign Policy', *Energy Policy*, Vol. 56, 2013, pp. 114–125.

28 'Medium Term Oil and Gas Markets, 2011', *International Energy Agency*, p. 248.

29 Kent A. Calder, *Asia's Deadly Triangle* (Nicholas Brealey Publishing, London, 1997), p. 155.

30 Lee Cordner, *Offshore Oil and Gas Safety and Security in the Asia Pacific: The Need for Regional Approaches to Managing Risks* (S. Rajaratnam School of International Studies, Monograph number 26, 2013), pp. 59–61.

31 Gurmeet Kanwal, 'IPI Pipeline Is a Good Option, but Security Is a Nightmare', *Gas and Oil*, 6 July 2008. http://www.gasandoil.com/news/2008/08/ntm83256. Accessed on 13 November 2013.

32 Robert Tuttle, 'Qatar's Gas Dominance Challenged', *Bloomberg News*, 1 April 2014.

33 'Saudi Arabia Oil Assets Immune to Terror Risk', *Reuters*, 28 September 2011.

34 Tom Parfitt, 'Terror Alert as Georgian Pipeline Opens', *The Guardian*, 28 May 2006.

35 *Economic Times*, 22 December 2011.

36 Sanjeev Choudhary, 'GAIL Will Deploy Drones to Guard Gas Pipelines to Raise Safety Standards', *Economic Times*, 20 October 2015.

37 Anne Korin, 'Why Energy Terrorism Is Nothing New and Hard to Stop', *World Politics Review Trend Lines*, 2 May 2013.

38 'Vandalism of Gas Pipelines and Power Generation', *This Day Live*, 15 February 2015. http://www.thisdaylive.com/articles/vandalism-of-gas-pipelines-and-power-generation/201723

39 The Yemeni minister of Oil and Minerals Hisham Sharaf said that the country lost approximately two billion dollars due to the frequent sabotage attacks targeting oil and gas pipelines, especially in the central province of Marib, *Yemen Post*, 28 May 2012.

40 Jad Mouawad, 'The Pipes Carry Clout with the Oil', *New York Times*, 14 May 2006.

41 Okechukwu Nnodim, 'Pipelines Vandalised 200 Times in Six Months', *Punch*, 12 February 2015. http://www.punchng.com/business/business-economy/pipelines-vandalised-200-times-in-six-months-fg/. Accessed on 3 January 2016.

42 Derek Brower and Kirk Sowell, 'Pipelines, Not LNG, for East Med Gas', *Petroleum Economist*, Vol. 81, no. 6, July–August 2014, pp. 26–27. Interestingly, it is Egypt that is now seeking the Israeli gas in a dramatic reversal of roles.

43 *AFP*, 10 June 2015.

44 The U.S. natural gas pipeline network, consisting of 210 pipeline systems, is a highly integrated transmission and distribution grid that can transport natural gas to and from nearly any location in the lower 48 states. http://www.eia.gov/pub/oil_gas/natural_gas/analysis_publications/ngpipeline/index.html. Accessed on 30 June 2015.

45 For the text of the Cyber Security Act, see https://www.govtrack.us/congress/bills/112/s2105/text. Accessed on 3 January 2016.

46 *Christian Science Monitor*, 5 May 2012.

47 Daniel Gaynor, 'Cybersecurity Next National Security Challenge', *San Francisco Chronicle*, 4 June 2012. http://www.sfgate.com/opinion/openforum/article/Cybersecurity-next-national-security-challenge-3606111.php

48 'Charter of the Pipeline Working Group and the Pipeline Sector Coordinating Council', November 2014. http://www.dhs.gov/sites/default/files/publications/Pipeline-SCC-Charter-508.pdf. Accessed on 3 January 2016.

49 'Cross-Border Oil and Gas Pipelines: Problems and Prospects', *Joint UNDP/World Bank Energy Sector Management Assistance Programme* (ESMAP), June 2003, p. xiii. http://siteresources.worldbank.org/INTOGMC/Resources/crossborderoilandgaspipelines.pdf. Accessed on 3 March 2014.

Part II

THE GAS TROIKA

3

IRAN

Gas pipelines under/after sanctions

Iran occupies a pivotal position on the tripolar chess-board. Geographically, it is the only nation that abuts both the Persian Gulf and the Caspian Sea, position-ing Tehran to play a significant role in the two areas of greatest energy concern to the United States, Rus-sia, and China. Iran also abuts the strategic Strait of Hormuz – the narrow waterway from the Gulf to the Indian Ocean through which about one-quarter of the world's oil moves every day. As a result, if Washington ever lifted its trade embargo on Iran, its territory could be used as the most obvious transit route for the deliv-ery of oil and natural gas from the Caspian countries to global markets, especially in Europe and Japan.

Above all, of course, Iran possesses the world's second largest reserves of petroleum – an estimated 132 billion barrels (11.1 percent of the world's known reservoirs); and also the second largest reserves of natural gas – 971 trillion cubic feet (15.3 percent of known reser-voirs). The Iranians may possess less oil than the Saudis and less gas than the Russians, but no other country con-trols so much of both of these vital resources.[1]

'Cooperation should certainly be carried out via Iran. For links between the north and the south, the east and the west, these countries and Europe, Europe and Asia, everything should cross Iran – oil and gas pipelines, rail-ways, communication routes and international airports', Rafsanjani to the Central Asian leaders.[2]

These two quotes bring out the centrality of Iran in terms of its geographi-cal location and energy assets as also the Iranian ambition to make the most of these advantages. They also identify the trade embargo as a factor that nullifies its privileged position on the global energy map. The chapter proposes to analyse Iran's gas endowments, its gas policies and some of the gas pipelines that it has attempted to propose, revive and build to break out of the shackles around its gas production, sale and purchase.

Iran's gas

Iran is the only country with vast reserves of oil and gas both. It is the second-largest oil-rich country after Saudi Arabia. And it is not the second-largest, but the largest gas-rich country in the world. In 2012, it overtook Russia as the largest depositor of gas reserves at 33.8 tcm or 18.2 per cent of the world's total proven gas reserves, according to the BP report.[3] The figure for Russia has stood at 31.3 tcm in comparison. Most of its gas is non-associated. Geographically, its gas fields are more scattered than the oil fields, but southern Iran dominates in gas reserves as is the case of oil reserves. Nearly 80 per cent of the gas fields are located there. The major gas fields are onshore fields of Kangan and Nar and the offshore fields of north and south Pars. Kangan gas is used for domestic consumption and reinjection.[4] Aghar–Dalan fields produce gas that is used for reinjection into the oil fields in Khuzistan. Khangiran is the oldest onshore field in the north, which is located near Sarakhs in the northeast and very close to Turkmenistan. Kangan and Khangiran are non-associated. A small field in Gorgan and a recent discovery of a field near Bandar Anzali on the Caspian Sea are in the north as well.

The pillar of Iran's gas programme is the gigantic offshore South Pars field on the Gulf, which is 300 km from Bushehr and 580 km from Bandar Abbas. A substantial part of its production will be exported as LNG. Tehran wants the Pars Special Econo-Energy Zone, established in 1998, to become 'one of the most important industrial energy poles of the Middle East'.[5]

For a long time, gas was used as a by-product of oil and most of it was flared. But in the 1990s, the need for reinjection of gas into fields has entirely reversed the dependency and now oil production depends on gas. Some of Iranian oil fields are close to maturity, and therefore, the country needs to reinject more than one-third of its gas into the oil fields to increase pressure and production.

A great deal of Iranian gas remains unexplored and the production is much below optimum. There are several reasons for this situation: the destroyed and damaged gas infrastructure during the Iran–Iraq war of 1980–1988, lack of technological up-gradation, poor investments, disputes with Saudi Arabia and Kuwait over the border through the northern Gulf continental shelf[6] and, of course, several sets of sanctions imposed by the

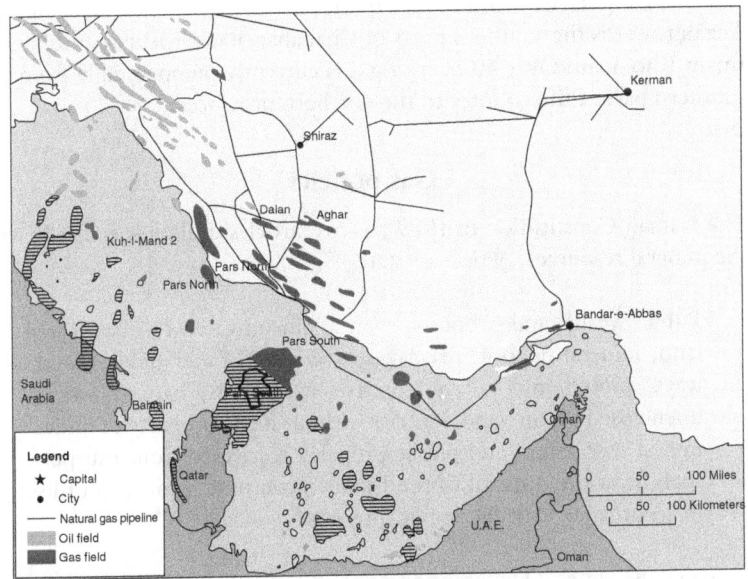

Figure 3.1 Oil and gas in Iran.

Source: Based on Energy Information Administration.

Disclaimer: Map not to scale.

United Nations, US and the European Union. On the positive side, the non/poor development of the gas fields also signifies a huge potential for future development.

Iran has one of the largest populations in the region. The domestic consumption of energy is constantly increasing. In addition, internal migration has created a peculiar condition of production and consumption divide. Through the long years of Iran–Iraq war, there was a steady internal migration from the South-west to the North-east. The North-east, as a result, developed much faster. Today, the urban centres, industries and agro-industries in the country are in the north around the Caspian rim. The gas, as noted, is largely concentrated in the south.

In the circumstances, there is a large domestic network of gas pipelines from the source to the consumers. Roughly 4,000 kilometres (km) in length, the network covers both the south–north and the east–west regions. Along a south–north axis, the Gulf is connected via pipelines to the Caspian Sea. The existing east-west pipelines extend from Sarakhs on the Turkmenistan border to Rezaieh near Turkey. The convenient transit routes, linking the landlocked Caspian Sea to the Gulf, are those via

Sarakhs, Dargas, Astra and the northern port cities of Neka, Noshahr and Anzali towards the southern ports of Chabahar, Bandar Abbas and Bandar Imam Khomeini. Over 40 bcm of gas is currently pumped daily from the southern part of the country to the northern provinces.[7]

Gas policies

The Iranian Constitution of 1989 has two articles dealing specifically with the mineral resources. Article 45 states:

> Public wealth and property, such as uncultivated or abandoned land, mineral deposit, sea, lakes, rivers, and other public water-ways, mountains, valleys, forests, marshlands, natural forests, unenclosed pastureland, legacies without heirs, property of under-mined ownership, and public property recovered from usurpers, shall be at the disposal of the Islamic government for it to utilise in accordance with the public interests.[8]

Article 81 states: 'The granting of concessions to foreigners for the for-mation of companies or institutions dealing with commerce, industry, agri-culture, services or mineral extraction, is absolutely forbidden.'[9] Together, the two articles prohibit private ownership of and foreign participation in the mineral resources and their development. On both counts, the com-mon resources are for common usage by the Iranian citizens exclusively.

Over time, some changes have been made on the issue. The 1993 budget allowed the National Iranian Oil Company (NIOC) to have contracts of up to $2.6 billion with competent foreign companies. The budget next year introduced terminology and mechanism of buy-back contracts as the legiti-mate source of attracting investment. The budget in 2003 authorised the NIOC to include both exploration and development in such contracts.[10]

As Iran gradually opened the doors to foreign participation in its energy sector, the foreign doors gradually shut in its face. The US imposed sanc-tions on its energy sector in 1996. The Iran Libya Sanctions Act (ILSA) of 1996 was renamed Iran Sanctions Act (ISA) in 2006. The United Nations and the European Union imposed their own sets of sanctions with the result that foreign oil and gas companies like the Anglo-Dutch Royal Dutch Shell, Spain's Rapsol and France's Total folded up and left Iran. Since the signing of the Joint Plan of Action between Iran and the P5+1 in Novem-ber 2013 and the Joint Comprehensive Plan of Action in July 2015, the country has been granted 'limited, targeted and reversible sanctions relief'.

The National Iranian Gas Company (NIGC) is responsible for natural gas infrastructure, transportation and distribution. The National Iranian

Gas Exports Company (NIGEC) was created in 2003 to manage and to supervise all gas pipelines and LNG projects. Until May 2010, NIGEC was under the control of the National Iranian Oil Company (NIOC), but the Petroleum Ministry transferred NIGEC, incorporating it under NIGC in an attempt to broaden responsibility for new natural gas projects.[11] In 2012, the NIGEC was liquidated by the Iranian Oil Ministry. The function of gas exports via pipelines was transferred to the state companies and the functions of commercial gas sales and export of liquefied gas were transferred to the international department of the oil company.[12] The frequent restructuring of the gas administration suggests a constant review of the gas policies. It also suggests Iranian responses to the tightening sanctions on its gas trade.

The most significant energy development project in Iran is the offshore South Pars field. It is one of the largest independent gas reservoirs in the world lying on the maritime border between Iran and Qatar in the Gulf. It is one of the country's main energy resources. This gas field covers an area of 9,700 sq. km, of which 3,700 sq. km belong to Iran. The Iranian portion is estimated to contain some 14 tcm of gas reserves and some 18 bb of gas condensates. This amounts to roughly 7.5 per cent of the world gas reserves and approximately half of Iran's gas reserves.[13]

The Pars Oil and Gas Company (POGC), a subsidiary of National Iranian Oil Company (NIOC), was established in 1998. POGC is responsible for development of all phases of South Pars gas field and development of North Pars, Golshan and Ferdowsi gas fields, as well. The main missions of POGC are reservoir evaluation, technical and economic assessments; engineering studies, executive contractors' appraisal and selection; and management of implementation of South Pars fields' development projects and other gas reservoirs in the Gulf.

Presently, some precise and sophisticated projects have been designed for development of 24 phases to produce 790 cm of gas per day. South Pars gas field development shall meet the growing demands of natural gas, injection into oil fields, gas and condensate export and also feedstock for petrochemical industries. As a result, Asalouyeh and Tombak ports, some 270 and 220 km south-east of Bushehr, respectively, have been selected as onshore locations for the construction of onshore installations of the phased development of this field.

Out of 24 phases, Phases 1–10 are online. The majority of South Pars natural gas development will be allocated to the domestic market for consumption and gas reinjection. The remainder will be either exported as liquefied natural gas (LNG) or used for gas to liquids (GTL) projects. A consortium led by GS E&C in Korea, which has teamed up with two Iranian contractors, the Iranian Offshore Engineering and Construction

Company and the Oil Industries Engineering & Construction Company, is to carry out the project. Phases 9 and 10 will produce natural gas ethane, gas condensates, LPG and sulphate.[14]

Iran made and pronounced a few technological feats in early 2000s. In 2002, it attained self-sufficiency in operations for transportation and installation of seabed pipelines at a depth of 100 metres with a capacity to lay 1.5 km of pipelines a day. The pipeline laying operation included 32-inch pipes from the refinery of Phase 1 of the South Pars to the onshore platform.[15] In 2005, a project to build five-decker platforms weighing five thousand tonnes each was announced in partnership with a foreign contractor that was to have 42 per cent of the shares with the Iranian contractors retaining 58 per cent of the shares of the expected cost of $150 m.[16]

Even more overly ambitious proposals followed. In 2008, it announced that it would be exporting the LNG via pipes; the Iranian part of the lines would extend to the border, while the rest of it would be made by the countries themselves.[17] The gas discovered on the Lavan Island in the Gulf would be used for exports in the LNG form.[18] Next year, it announced a plan to increase the length of the high-pressure gas pipelines from 30,000 km to 70,000 km by 2025.[19] In 2012, it claimed that it would soon be implementing gas-to-wire project under which it would export electricity instead of gas to the countries in its neighbourhood.[20] In 2014, it announced that Shourijeh, an underground gas storage facility in Iran's north-eastern province of North Khorasan, would soon be operational. It was expected to hold as much as 700 mcf (20 cm) gas for Iran. With this, Iran 'will become one of five main countries in the world and second in the Middle East with underground facilities for natural gas storage', the managing director of Natural Gas Storage Company Massoud Samivand claimed.[21]

With the accomplishments came the acknowledgements of the limitations. In 2010, it decided to drop its plan to develop the LNG and concentrate on developing the pipelines further. The decision was prompted by the departure of the foreign companies that were managing the Phases 13 and 14 of the South Pars gas fields. Multiple reasons were extended to justify the reversal: Iran has a large network of pipelines; it has long borders and friendly neighbours as its piped gas customers; it is cheaper and faster to export gas via the pipes; the LNG technology is expensive, complicated and time-consuming; and it demands huge investments, according to NIOC's managing director Ahmad Ghalebani.[22] Iran's refining capacity was hurt due to the sanctions. The cars were mandated to have a dual-fuel capacity – to run on gasoline and on compressed natural gas (CNG). The CNG was cheap; the gasoline was more costly and also rationed.

Caspian Sea[23] is a major focus of Iran's energy policies.[24] Initially, Iran and Russia favoured the inner sea concept, and the rest insisted on treating

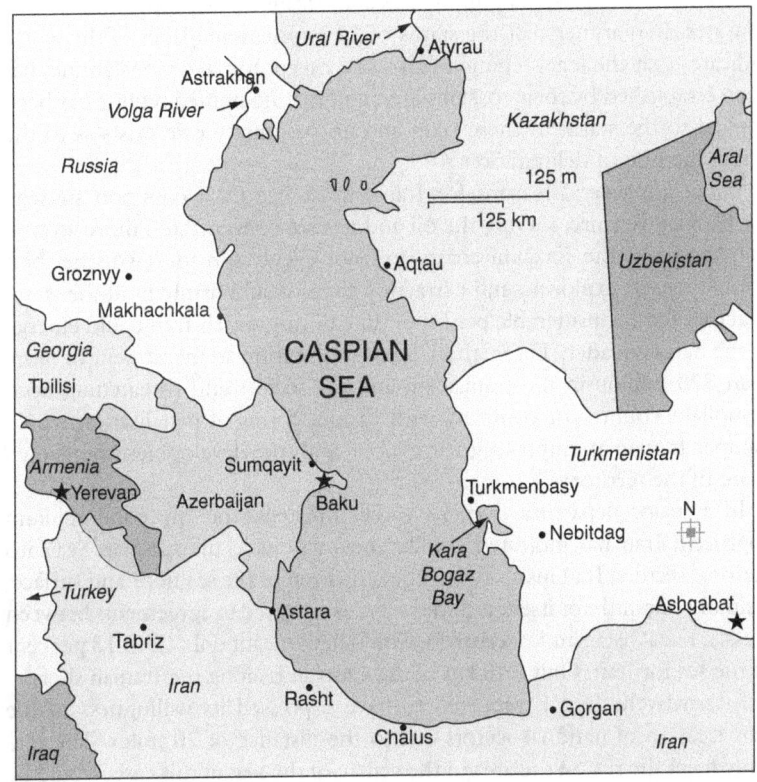

Figure 3.2 The Caspian Sea.

Source: Based on http://www.worldatlas.com/aatlas/infopafe/caspiansea.htm

Disclaimer: Map not to scale.

it like any other sea. Under the first option, the underwater resources would be jointly developed and equally shared among the coastal states; under the second option, the law of the sea would apply according an exclusive economic zone to each.[25] Russia broke ranks with Iran in 1996. In May 2003, Azerbaijan, Kazakhstan and Russia concluded a trilateral agreement, putting in place a de facto legal regime in the northern Caspian.

Iran continues to reject the law of the sea regime for the Caspian. It maintains that the Caspian is not a sea at all, but an inner sea or lake. The law of the sea does not apply to the Caspian, as per this position. The Caspian Sea is a landlocked body of water with no natural link to the world's oceans and seas. Therefore, according to Iran, norms and principles of international law governing gulfs, seas and oceans do not apply to this body of water. Iran argues that the status of 27 international lakes of the world indicates that the legal regime of all lakes, except for lake Constantine, has been established by their coastal states, and that no unified regime has been applied to the status of these lakes and no customary rule exists as to the actual method of delimitation.[26]

There are several reasons for Iran's insistence on a common strategy for Caspian resources. One, the oil and gas are concentrated more around the Azeri and the Kazakhi coasts and not evenly spread. Two, the joint effort towards exploring and extracting them would firmly bind the states together for a considerable period of time during which Iran could emerge as the natural leader. Three, the US law prohibiting an investment of more than $20 million in the Iranian oil and gas sectors and threatening non-compliant country or company with various forms of penalisation would hamper Iranian attempts to go at it alone with the development of its own share of the territory.

In a major departure from its earlier insistence on the condominium approach, Iran has indicated a willingness to divide the Caspian Sea into national sectors. Iran insists on an equal division of the sea floor and surface, claiming one-fifth for itself. Equidistance, as accepted in agreements between Russia, Kazakhstan and Azerbaijan, would have meant only 12 to 13 per cent of the sea for Iran. Only 740 km of the Caspian is along the Iranian shore.

Alternatively, Iran is reported to have expressed its willingness to the demarcation of national sectors out to the distance of 20 miles. The revenue from the rest of the area in the centre of the sea, in this case, would be divided equally among the five littoral states. The proposal has since been named the 'doughnut plan', as the area would resemble a doughnut with a hole in the middle. Iran rejects the Russian proposal in this regard that would demarcate the national sectors up to 45 miles. Its own proposal of extending the national sectors only up to 20 miles would leave a large common area for joint development.

For a while, there was a talk of 'cakewise plan', under which each state would have got a slice of the waterfront based on the length of its shoreline. Since Iran's shoreline does not have rich deposits of oil and gas, the plan was not well received in Tehran.

As the Azeris and the Kazakhis went on with the development of their sectors, Iran followed suit.[27] Accordingly, Khazar Exploration and Production Company was formed and affiliated to the National Iranian Oil Company. On 15 December 1998, it entered into a major deal with the Anglo-Dutch Shell and the UK independent Lasmo to carry out a study of 10,000 sq. km. of unexplored waters in the Iranian sector of the Caspian.[28] According to the US Department of Energy, Iran had already explored and identified 40 reservoirs containing as much as 3 bb of oil in its territorial waters in the Caspian.[29]

In the final analysis, the Caspian dispute has always been about oil and gas rather than the water. As the former Armenian president Robert Kocharian had famously remarked at the World Economic Forum in Davos: 'Is there any water in the Caspian, or is it only oil?'[30] With time, the issue of foreign military presence has overtaken the energy aspect of the dispute. The Caspian Summits have managed to retain a consensus on keeping the NATO forces out of the Caspian waters. The Fourth Caspian Summit held in Astrakhan in September 2014 has left it to the Fifth Summit to be held in Astana to finalise the legal status of the Sea. The Iranian Defence Minister Hossein Dahqan has reiterated that the regime of the Caspian Sea 'can be perfectly resolved through negotiations'.[31] The dates for the Summit are not yet finalised.

Pipelines from Iran: successes and failures

Iran was the first country in West Asia to export gas by pipeline. In the mid-1970s, Iran's gas network was connected to the Soviet network in Azerbaijan and large volumes of gas were exported to the Soviet Union. Iran has always aspired to be at the centre of global energy. Straddling two energy rich areas of Central Asia and the Gulf, it imagines itself as the 'Golden Gate between the Caspian and the Gulf'. An Iranian energy analyst claims that the country can rightly be termed the world's 'Energy Bridge' constituting the main artery link between the Caspian littoral states and the West, South and the East.[32] Considering the geography and also the geology, it should be an eminently reasonable aspiration.

The Institute for Political and International Studies (IPIS), the apex research think tank in Tehran, developed Project PEACE (Pipeline Extending from the Asian Countries to Europe) in the nineties to give a concrete shape to the aspiration. It proposed a gas loop system within Iran with more than half of it already in existence. The major participants in the project were identified as the World Bank and other financial institutions; major gas producers like Turkmenistan, Uzbekistan, Kazakhstan, Iran, Iraq, Kuwait, Qatar, Saudi Arabia, UAE and Oman; and major gas consumers like Azerbaijan, Armenia, Georgia, Ukraine, Turkey, Pakistan, India and Afghanistan.[33] It was made explicit that most of the projects would either pass through Iran or its territorial waters.[34]

The country would soon find itself under sanctions, which would continually become tighter and expand to newer areas of economic and financial spheres. In the circumstances, the Iranian initiatives towards building pipelines beyond its national borders have not always been successful.

An early success with the Iran–Armenian pipeline

Iran's gas pipeline with Armenia is an outstanding example of collaboration between two neighbours; both under sanctions from different sources. Turkey and Azerbaijan have imposed economic sanctions over the ethnic Armenian enclave of Nogorno–Karabakh inside Azerbaijan. Armenia is almost totally dependent on the Russian Gazprom for its gas requirement, and the agreement with Iran was to reduce its reliance on Russia. It was signed in 1992. The pipeline was planned to run 101 km in Iran from Tabrez to the Armenian border town of Meghri and another 41 km from there to Kajaran. From there, it would reach central Armenia.

The Chinese were the first to evince an interest in participating in the venture. Representatives from a Chinese company arrived in Yerevan in April 2000 to hold talks with the Armenian side on the issue.[35] In 2001, the EU stepped into the act with financial contribution. Sebastian Dubost, the programme coordinator for the EU's representative office in Yerevan, said in January 2001 that Brussels had approved plans for the Iran–Armenia pipeline and would make the sum of Euro 30 million available for the drafting of the technical and legal documents needed to launch construction. Then in February, the EU announced that it had decided to grant Armenia a sum of Euro 3 million to facilitate the formation of a consortium to build the pipeline.[36] In October, the EU was reported to have funded a $10 million feasibility study for pumping Iranian gas through the Greek connection.[37]

The EU support made the project feasible and attractive. Greece joined the project in March. During a meeting in Athens, the Iranian, Armenian and Greek officials signed a Memorandum of Understanding.[38] They acknowledged a Greek engineering company's work on a feasibility study for the pipeline. Iran's national gas company and Gaz de France were authorised to set up a consortium to build the conduit.

The Armenian president Kocharian paid an official visit to Iran in 2001, signing accords over cooperation in the fields of civil engineering, energy, environment, customs and protection of historic and cultural monuments.

Ukraine, in a bid to reduce its gas dependence on Russia – like Armenia, showed interest in joining the project at its very initial stage. The Ukrainian foreign minister Anatoly Zlatko visited Tehran in December 2001 and at a joint press conference with his Iranian counterpart Kamal Kharrazi called for the transport of gas to Western Europe via Ukraine.[39] Poland expressed its wish to join the project in February 2002.[40] Russia entered into the picture soon thereafter. Gazprom purchased a majority share in the Armenian section of the pipeline through its subsidiary Armosgazprom. There were rumours that the diameter of the pipe was reduced from 1,420 to 700 millimetres under pressure from Gazprom, so that Iran could not export its gas to Europe and compete with the Russian share of the European market.[41]

In 2007, the first section of the line was opened at the border in the presence of Mahmoud Ahmadinejad and Robert Kocharian, the presidents of Iran and Armenia, respectively.[42] Yerevan was to pay for the gas with electricity it produces at a Soviet era nuclear power plant. The pipeline is operational since 2009. Armenia has since been asking to enhance the gas supply.

A fairy tale Iran–Turkey pipeline

On 12 August 1996, Turkish prime minister Necmettin Erbakan signed a $20 billion natural gas deal with Iran. Turkey was supposed to start importing Iranian gas by 1999 for 23 years. The deal was strongly opposed by the Bill Clinton administration as it was a clear violation of the ILSA. It sent a delegation to Ankara to dissuade Turkey from going ahead with it. Erbakan refused, arguing that he was implementing a long-established policy of diversifying its energy resources. It was an example of the geopolitics trumping all other considerations, including Turkey's membership in the North Atlantic Treaty Alliance (NATO) and its close ties with Washington.[43]

In order to avoid US sanctions, each country was responsible only for work on the pipeline portion lying on its own soil. The gas transportation costs for Turkey were lowered because the pipeline use, from the Iranian side, was shared with the gas delivered to Iranian users of the northern provinces. The gas flow was postponed several times because of work delays on both sides. Reportedly, US opposition also seemed to have played some role in the delays by blocking the delivery of powerful compressors for the project. In the end, in December 2001, Turkey started importing gas from Iran via a 2,577-km pipeline from Tabriz to Ankara.[44] There was a two-year delay over technical details and bilateral wrangling before the actual delivery began.[45]

The project was mired into complications much beyond the bilateral problems. It was clearly a violation of the US ILSA. Turkey's role within the NATO and its importance for the US policies in the region made it immune to any punishment that a non-compliance with ILSA would have stipulated. The Armenian lobby inside the US political system would have demanded a strong retribution. The Turkey–Israel relations would have been damaged. And Turkey was into the deal very openly, not willing to resort to covert methods. Was containing Iran more important or protecting an ally?

A fairy tale, therefore, was spun around the entire project. It was publicised as the Turkmen gas passing through a short transit line across Iran via a physical swap. There was no such line that traversed the length of Iran from Turkmenistan to Turkey, and the relevant politicians and the policy-makers knew it. Even the most vociferous adversaries seemed to acquiesce in the story. An observer described it as the 'Phantom Pipeline'.[46]

Turkey was a prime purchaser of natural gas in those years. Two giant projects, the Blue Stream from Russia[47] and the Trans-Caspian from Turkmenistan were racing to reach the Turkish market first. Turkey itself had signed various deals of varying degrees of firmness to purchase gas from Russia, Turkmenistan, Azerbaijan and Iraq. Turkey hoped to use most of the imported gas and also re-export the rest to Europe. Iran, on the other hand, was looking beyond Turkey to Greece and thence to Europe.

The Turkish economy registered a downturn in early 2000s. The estimates of gas requirement were found to be inflated. Within six months after the first delivery, Turkey cut off gas imports from Iran. The Turkish Energy Minister Zeki Cakan blamed the quality of Iranian gas, but Iran's oil minister, Bijan Namdar Zanganeh, argued that the real reason was the slumping Turkish economy. He even suggested that Turkey had halted its imports from Iran because Russia had undercut them by offering a lower price, according to the official Iranian news agency IRNA.[48] Zanganeh threatened to seek penalties against Turkey under the terms of a 1996 take-or-pay contract that was supposed to guarantee more than $20 billion in Iranian gas sales over 25 years. In November 2002, the gas flow from Iran to Turkey was resumed, after negotiating a lower gas price.

Breaching the GCC unity through the Iran–Oman pipeline

Oman's ties with Iran owe a lot to Iran's military support to Sultan Qaboos in his fight against the insurgency in Dhofar from the early to mid-1970s. Qaboos's personal appreciation and Oman's state interests converge, when it comes to the bilateral relations. Underpinning the common stands on critical issues are solid economic reasons. There has been a sharp increase in demand for oil and gas in Oman as the government investment in infrastructure and industrial development is going up. Oman receives 18 per cent of its gas imports from the Dolphin project of Qatar. Qatar has put a moratorium on its gas export,[49] and Oman must look for other sources. Oman and Iran have agreed to develop gas in the Kish Island of Iran and transport the same to Oman via a 200-km long pipeline under the Gulf waters.[50] Oman has agreed to bear the entire cost of $12 billion. A joint development of Hengam/Bukha gas field at the cost of $200 million and transporting it to Oman via a 100-km pipeline is also agreed.

For Iran, Oman is a friend across the Gulf water and an economically. As the sanctions regime tightens around Iran, any opening that offers itself must surely be welcomed. Oman is one such. In addition, Oman offers a window to reach out to wider world. For example, the LNG plant at Qalhat in Oman has a small capacity of just under 10 million tonnes a year. There are plans to expand the same, which could be a future transit route for Iranian gas.[51]

The Iranian president Hassan Rouhani visited Oman in March 2014. A 'Heads of Agreement' was finalised during the visit. Gas from Iran may arrive as early as 2017, Oman's oil minister Mohammed Al-Rumhy said. Oman will pay for the pipeline, which will extend from the Iranian Province of Hormuzgan to Sohar in Oman, and some of the gas may be re-exported to neighbouring countries, his Iranian counterpart Bijan Namdar Zanganeh added.[52]

The pipeline is significant in a number of ways. One, Iran will now have the possibility of converting its gas to LNG at the Omani liquefaction facility and re-export the same.[53] Till now, Iran did not have financing and requisite technology to develop the liquefaction independently. Two, Oman is the second country after Turkey to have defied the sanctions imposed on Iran. Three, Oman has also annoyed the GCC of which it is a member by extending its energy ties with Iran rather than with the GCC states. Four, it is the first project that Iran has signed with a GCC country.

A stillborn Iran–Iraq–Syria pipeline

Chapter 1 details the discoveries of immense energy reserves in the eastern Mediterranean; both the Levant basin located along the shores of Syria, Lebanon, Israel, Gaza and Cyprus and the Nile basin north of Egypt. Syria alone is estimated to have discovered proven gas reserves of 284 bcm, oil reserves of 2.5 bb and shale reserves of 50 billion tonnes with the possibility of more findings.[54]

Syria became an indispensable link in two competing pipeline proposals floated around this time: The Iran–Iraq–Syria–Lebanon pipeline and the Qatar–Saudi Arabia–Jordan–Syria–Turkey pipeline. The Iran–Iraq–Syria–Lebanon pipeline was dead at birth.[55] A larger context of the failed initiative would bring in an earlier proposal from Qatar. Qatar has the third largest reserves of gas after Iran and Russia. Estimated at 25 tcm, most of its gas exports are in the form of LNG. The shale gas production in the US will impact the sale of Qatari LNG; therefore, Qatar seeks to secure long-term contracts via pipelines to the European countries. The EU has secured its energy imports till 2030 and is looking for secure infrastructural investments for the future thereafter. The Nabucco pipeline project[56] from eastern Turkey to Austria is stalled due to insufficient gas available.

It was in this context that a new pipeline for Qatari gas was proposed. In 2009, during the Qatari Emir Sheikh Hamad bin Thani's visit to Turkey, it was agreed to build a pipeline and link it up with the Nabucco in Turkey.[57] It was to originate in Qatar and move through Saudi Arabia, Jordan and Syria, reaching Turkey. The European markets would share the resource with Turkey.

Qatar would not be the sole beneficiary of the pipeline. There were three distinct additional calculations behind the Qatari proposal. It would pry Turkey loose from dependence on Iranian supplies, it would severely curtail the Russian near-monopoly as the sole gas supplier to Europe and it would facilitate Israel's gas export to Europe. The West has high stakes too and not just to contain Russia. Europe has been struggling to establish an ambitious 'southern energy corridor' through a combination of pipelines that would transport a huge amount of 60 to 120 bcm per year of West Asian and Caspian energy to Europe. The Qatari gas is an indispensable component in the success of the venture – the gas that would traverse through Qatar–Saudi Arabia–Jordan–Syria–Turkey.

It was against this background that a Memorandum of Understanding for the Iran–Iraq–Syria–Lebanon pipeline was signed in the Iranian port city of Bushehr on 25 June 2011. It was agreed to construct a gas pipeline from the Iranian gas field of Assaluyeh through Iraq and Syria.[58] To be built at a cost of $10 billion, its projected capacity of 110 cm per day was tentatively allocated among Iraq, Syria and Lebanon. It was proposed to extend it to Greece through a submarine line and from there on to the markets in Europe. Named the 'Friendship Pipeline',[59] it was to be supplemented by the export of LNG from the Syrian ports on the Mediterranean. Latakia and Tartous are two major Syrian ports. Russia has leased Tartous and constructed a naval base there.

There are many interesting conspiracy theories making rounds around this pipeline.[60] One sees a clear connection in the timing between the signing of the memorandum on the Iran–Iraq–Syria gas pipeline and the beginning of violent uprising in Syria. The other notes a coincidence between the fiercest fighting in places inside Syria and the proposed route of the Qatari pipeline through its territory. Yet another explains the Qatari support for the Muslim Brotherhood among the Syrian rebels and beyond in the region in this context. After all, Qatar has sunk $3 billion in the Syrian civil war. A small sum in Qatar's wealth, but big in comparison to the Western handouts to the rebels. An additional consideration could be that Qatar shares its gas field, which it calls the North Dome with Iran, which calls it

South Pars. Together it is the largest gas field in the world. The disputes of the past may flare up in future over the borders and extraction rights in the gas field. The pro-Russian voices speculated that Qatar was seeking to over-throw Assad with a view to squeeze Russia out of its traditional markets.[61]

On 31 July 2013, Prince Bandar of Saudi Arabia made a hurried visit to Kremlin. He was the director general of the Saudi Intelligence Agency at that time. He is reported to have pleaded his case for regime change in Syria and offered some incentives to Putin by way of $15 billion worth of arms contracts, security against terror attacks from Saudi-funded Chechens during the 2014 Winter Olympic Games to be played in Sochi and more. He is also believed to have offered an assurance that whatever regime came after Assad, the Saudis would not sign any contracts damaging Russian interests by allowing Gulf countries to transport their gas across Syria to Europe. Yet another conspiracy theory?[62] The threats and incentives from Bandar to Putin were leaked and published in the Russian press. The source of the leaks could only have been Putin's office.

In the meanwhile, Iran is plodding on – as if totally indifferent to the reality on the ground. In late November 2012 the governor of Gilan-e Gharb Province Ardeshir Rostami called an executive co-ordination meet-ing of the officials to announce that the construction of the pipeline had reached from Kuhdasht to the Gilan-e Gharb to cross the border en route to Baghdad. He said that the pipeline was to be 225 km in length and would be completed by June 2013 with an investment of $3 billion.[63] The Iraq deputy prime minister Hussain al-Shahristani also expressed hope that the line could be operational by November 2013.[64] The Fars News Agency reported in September 2014 that the line to the Iraqi power plant in Man-soureh was complete. The Iranian gas along the 48-inch-diameter pipeline would be exporting 25 cm a day, though it would begin by delivering 7 cm for the present.[65]

In the most optimistic scenario, the pipeline may end at the Iraqi power plant in the face of the turbulence in Syria.

Resurrecting Nabucco?

The sanctions on Iran are already unravelling. The Iranians know this and are all set to open up the gas spigot in full force. Ali Majedi, Iran's deputy oil minister for international affairs, has been the first to voice a welcome to Europe as a major recipient of Iranian gas bonanza.[66]

Speaking on the sidelines of the Middle East Petroleum and Gas Con-ference in April 2014, he identified three passages through which the locked-up Iranian gas could move outward: through Turkey, through Iraq–Syria–Lebanon or through Armenia–Georgia–Black Sea. 'The Turkey pipeline would be the cheapest way,' he said. Discretion was forgotten in his surge of new-found confidence as he went on to chide Russia. Iran and Russia would find it hard to agree on any oil and gas deal, because they are rivals in both markets. 'There is no way that Iran will receive some of the oil from Russia. Maybe vice-versa, maybe. But not now,' he said. It was an oblique reference to sanctions on Russia that may one day require the country to import oil from Iran. Mohammad Reza Nematzadeh, the Iranian minister of industries, attempted a weak exercise in damage control after Majedi's comments, when he added that 'We don't want to compete with Russia. But we know that Europe's demand for gas is increasing and we would like a share in this'.[67] His statement came almost immediately after Majedi's comments. Aspiring for a share of Russia-dominated gas market in Europe was only slightly less of a threat than Majedi's provocation!

The pro-Russian voices, in the meanwhile, are panicky.

> And why is Iran climbing into bed with the Great Satan instead of pushing for their original plan which was to build a pipeline from Iran-Iraq-Syria? Could it be that Uncle Sam is going to make sure that that pipeline never gets built? Is that why Washington is letting a couple thousand homicidal maniacs (ISIS) run around Syria and Iraq lopping off heads and wreaking havoc; so it will be impossible to lay pipe or transit gas? It sure looks that way.[68]

The next to voice the welcome has been Iran's oil minister, Bijan Namdar Zanganeh. In an interview with IRNA, he confirmed that when the South Pars gas field comes on stream, Iran would have extra gas to export. He said his country was planning on exporting 80 billion cubic metres of gas to Turkey, Europe, Iraq, Pakistan and the Gulf States. Shipments to Europe could be via a pipeline or in the form of LNG.[69] He, thus, put a definite figure on the exportable gas and expanded the list of the prospective buyers to non-European states.

Up the chain of command, the Iranian president Hassan Rouhani brought up the offer in a meeting with Austrian counterpart Heinz Fischer in September on the sidelines of the UN General Assembly. Rouhani told Fischer that 'the Islamic Republic can be a reliable supplier of energy for Europe' and specifically mentioned the Nabucco pipeline.[70] The choice of Austria to mention Nabucco is significant, as the Nabucco pipeline was originally planned to run from Erzurum in Turkey and end at Baumgarten an der March in Austria and the project was spearheaded by Austria's OMV.

Later, during his state visit to Vienna, he publicly stated that his country could assure amounts up to 25 bcm per year to Austria.[71] At the same time, he sought not to alienate Russia beyond a point. In an interview with the Russian TV Channel Rossiya 1, he introduced two caveats: one, 'We are lagging in production and think about domestic consumption first', and two, 'our production is far from this stage yet', when Iran can supply gas if Russia stops supplies tomorrow.[72] Rouhani's statement in Vienna and his interview on the Russian television are not necessarily contradictory. His commitment to supply 25 bcm per year to Austria did not need to be delivered immediately, and his inability to replace Russia as a gas supplier did not have to happen 'tomorrow'.

The revival of Nabucco was reported to be discussed in detail in Sofia, when the Iranian ambassador to Bulgaria Abdollah Norouzi met Prime Minister Boyko Borisov. Also present in the meeting were Bulgaria's deputy prime minister in-charge of absorption of EU funds and economic policy Tomislav Donchev, Foreign Minister Daniel Mitov and Energy Minister Temenuzhka Petkova. The meeting was held behind closed doors. After the meeting, Norouzi said, 'One of the means of diversification of gas supplies in Europe is the use of the Nabucco gas pipeline with the partnership of the Islamic Republic of Iran.'[73] Note that it was Bulgaria's withdrawal which led to the collapse of the Russia-initiated South Stream pipeline. The country obviously looked forward to the revival of the West-initiated Nabucco. The Iranian engagement with Bulgaria on the issue of Nabucco was an indirect slight to Russia.

Talks over Iran's participation in Nabucco coincide with a parallel project which is pursued by the Swiss energy group EGL.[74] A package of proposal worth 18 to 22 billion Euros was signed between the EGL and the NIOC in 2008. The scheme was planned to take the Iranian gas through Turkey, Greece, Albania and Italy to Switzerland.[75] The offer met with opposition in the Swiss parliament. It was decided to keep the contract non-operational, but it has not been cancelled.[76] There are moves to resurrect the scheme.

The European Union is quietly increasing urgency of the plan to import gas from Iran as relations with Tehran thaw while the relations with Moscow grow chillier, according to an unnamed EU source.[77] Richard Morningstar, US envoy for Eurasian Energy, has refused to rule out Iranian involvement in Nabucco, and suggested that opening up the Iranian energy sector could be a 'carrot' for improving relations on other issues.[78]

There seems to be much more going on than meets the eye. Preparations seem to be in full swing involving not just the potential buyers in Europe, but also the transit countries like Turkey. Martha Brill Olcott, a leading expert on Central Asia and the Caspian, may still come true in her prediction that Nabucco may, in the future, rise from its grave, quite possibly without Turkmen participation.[79]

Recapitulation and leads

Contrary to the popular perceptions, it is Iran and not Russia that possesses the world's largest reserves of gas. It also holds the second-largest reserves of oil after Saudi Arabia. Its location abetting the energy-rich Gulf and the Caspian and the energy passage the Strait of Hormuz gives it a unique place on the global energy map. At the same time, various sets of sanctions imposed on it by the UN, the US and the EU have stunted its energy production and commerce in multiple ways. The Iranian Constitution prohibits private ownership of and foreign participation in the mineral resources and their development although there have been gradual relaxations opening up the energy sector.

Iran's initiatives on gas pipelines have not always been successful. One of the earliest successes has been a short distance line running to Armenia, which has been functioning since 2007. It is an interesting example of two countries under sanctions – Iran from the US and Armenia from Turkey and Azerbaijan – coming together. Another short distance line runs to Turkey. Everyone seems to be in denial of this line as it violates the US ISA and is opposed by Israel and the Armenian lobby in the US. Turkey's close ties with the US and its membership in the NATO has tipped the scale on the matter. A lie has been created that it is the Turkmen gas that is transiting through Iran; although there is no line passing from Turkmenistan to Turkey. Variously, it is called a fairy tale or a phantom pipeline. Iran has finalised a 'Heads of Agreement' with Oman to deliver gas by 2017. The deal is significant on three counts: one, Oman is the second country after Turkey to have defied the sanctions imposed on Iran. Two, Oman has also annoyed the GCC of which it is a member. Three, it is the first project that Iran has signed with a GCC country. There is strong perception that sees a clear connection between a Memorandum of Understanding between Iran, Iraq and Syria in June 2011 to build a gas pipeline and the beginning of violent uprising in Syria.

Today, the sanctions on Iran are already unravelling. The Iranians know this and are all set to open up the gas spigot in full force. There is hope that the Nabucco pipeline – long in hibernation – may be resurrected with gas inputs from Iran.

Notes

1 Michael T. Klare and Tom Engelhardt, 'The Tripolar Chessboard', *Tom Dispatch*, 16 June 2006.
2 *Ittelat*, 18 February 1992; *FBIS*, 18 February 1992. Quoted in David Menashri, *Central Asia Meets the Middle East* (Frank Cass, London, 1998), p. 84.

3 The Russian reserves stood at 31.3 tcm *British Petroleum*, 2014. http://www.bp.com/en/global/corporate/about-bp/energy-economics/statis tical-review-of-world-energy/review-by-energy-type/natural-gas/natural-gas-reserves.html. Accessed on 21 October 2013. Also, *Asharq al-Awsat*, 19 June 2014. http://www.aawsat.net/2014/06/article55333417/iran-still-holds-worlds-largest-gas-reserves-bp. Accessed on 21 October 2013. For reserves world-wide, see Chapter 1.

4 Shah Alam, 'Iran's Hydrocarbon Profile: Production, Trade and Trend', *Strategic Analysis*, Vol. 25, no. 1, April 2001, pp. 119–134.

5 Pepe Escobar, 'Iran Takes over Pipelineistan', *Asia Times*, 10 September 2005.

6 In 2000, Saudi Arabia and Kuwait signed a bilateral agreement dividing the disputed Dorra gas field after Iran began drilling there. *Dolphin Company*. http://www.dolphinshipping.net/offshoreprojects.html. Accessed on 13 June 2013.

7 Paola Ceragioli and Maurizio Martellini, 'The Geopolitics of Pipelines', *Asia Times*, 29 May 2003.

8 *The Constitution of the Islamic Republic of Iran* (Islamic Propaganda Organisation, Tehran, 1990), p. 45.

9 Ibid, p. 58.

10 Nima Nasrollahi Shahri, 'The Petroleum Legal Framework of Iran: History, Trends and the Way Forward', *China and Eurasia Forum Quarterly*, Vol. 8, no. 1, Spring 2010, pp. 121–122.

11 *Energy Information Administration*, November 2011. http://www.eia.gov/cabs/Iran/Full.html. Accessed on 21 October 2014.

12 *Mehr News Agency*. Quoted in the *Journal of Turkish Weekly*, 14 July 2012.

13 *Pars Oil and Gas Company*. http://www.pogc.ir/Default.aspx?tabid=136. Accessed on 13 May 2015. The Iranian estimates are considered inflated. See Chapter 1.

14 'Iran Completes Development Plan for South Pars Phases 9 and 10 by 84%', 24 February 2008. http://www.gasandoil.com/news/2008/03/ntm81210. Accessed on 3 March 2013.

15 'Iran Now Self-Sufficient in Installing Seabed Pipelines', *Asia in Focus*, 2 June 2002.

16 16 August 2005. http://www.gasandoil.com/news/2005/08/cnm53520. Accessed on 3 March 2013.

17 http://www.gasandoil.com/news/2008/08/ntm83239. Accessed on 3 March 2013.

18 26 October 2008. http://www.gasandoil.com/news/2008/11/cnm84889. Accessed on 3 March 2013.

19 http://www.gasandoil.com/news/2009/08/cnm93321. Accessed on 3 March 2013.

20 *Press TV*, 23 June 2012.

21 Daniel Graeber, 'Iran to Expand Gas Storage Facilities', *UPI*, 31 March 2014. Also see 11 April 2011. http://www.gasandoil.com/news/2011/04/iran-opens-1st-gas-storage-facility-in-middle-east. Accessed on 3 March 2013.

22 9 August 2010. http://www.gasandoil.com/news/2010/11/cnm104462. Accessed on 3 March 2013.

23 Caspian Sea is 1,200 km long; its width in the largest part is 550 km, and it occupies an area of about 370,000 to 400,000 sq. km. EIA estimates

48 billion barrels of oil and 292 trillion cubic feet (8.3 tcm) of natural gas in proved and probable reserves in the Caspian basins. Almost 75 per cent of oil and 67 per cent of natural gas reserves are located within 100 miles of the coast.

24 An internal sea between Russia and Persia, the 1921 Treaty of Moscow, which was reaffirmed in 1935, referred to it as a 'Soviet and Persian Sea'. After the Soviet disintegration three more states – Kazakhstan, Azerbaijan and Turkmenistan – share the Caspian littoral.

25 Under the United Nations Law of the Sea Convention (UNCLOS), if the Caspian is a legal 'sea', each littoral state receives a territorial sea up to 12 nautical miles, an exclusive economic zone (EEZ) up to 200 nautical, and a continental shelf. Since the Caspian at its widest is less than 200 miles, UNCLOS dictates the states apply a median line between claimants.

26 Djamchid Momtaz and Saeid Mirzaee Yengejeh, 'The Legal Regime of the Caspian Sea: Iranian Perspectives', *Iranian Journal of International Affairs* (Tehran), Vol. 13, no. 2–3, Summer–Fall 2001, pp. 237–238.

27 'Otherwise, the nouvo riche Government of Azerbaijan with the Help of Its American and Zionist Allies, Will Finally Gobble Up the Entire Undersea Crude', *Iran News* (Tehran), 12 December 1998.

28 The deal was worth $19.8 million, that is, just under the threshold of $20 million permitted under the US-imposed sanctions.

29 Geoffrey Kemp and Robert E. Harkavy, *Strategic Geography and the Changing Middle East* (Brookings Institution Press, Washington, DC, 1997), p. 103.

30 *Washington Times*, 6 February 2001. Quoted in Alec Rasizade, 'The Caspian Energy Legend and the 'Great Game' of Concomitant Pipelines', *Iranian Journal of International Affairs*, Vol. 14, no. 1–2, Spring–Summer 2002, p. 27.

31 *Tehran Times*, 21 April 2015.

32 Mohammad Sarir, 'Utilisation of Oil and Gas in the Caspian Region', *Amu Darya* (Tehran), Vol. 2, no. 1, Spring–Summer 1997, p. 7.

33 Narsi Ghorban and Mohammad Sarir, 'Oil and Gas: An Outlook for Future Cooperation among the Persian Gulf States', *Iranian Journal of International Affairs* (Tehran), Vol. 5, no. 3–4, Fall–Winter 1993–94, pp. 750–751.

See also Narsi Ghorban, 'Iran's Potential Role in the Development and Utilisation of Oil and Gas of the Caspian Region', *Iranian Journal of International Affairs*, Vol. 12, no. 2, Summer 2000, pp. 264–273; and Narsi Ghorban, 'Oil and Gas Pipelines from the Caspian Basin', *Amu Darya* (Tehran), Vol. 4, no. 1, Spring 1999, pp. 27–39.

34 Narsi Ghorban, 'The Evaluation of Recent Gas Export Pipeline Proposals in the Middle East', *Iranian Journal of International Affairs*, Vol. 7, no. 2, Summer 1995, p. 450.

35 'Chinese Firm Mulls Joining Iran – Armenia Gas Pipeline Project', *BBC Monitoring*. Source: Snark News Agency, Yerevan, 11 April 2000.

36 'EU Grants Armenia EUR3M to Start Gas Pipeline – Official', *Dow Jones International News*, 16 February 2001.

37 *Financial Times*, 21 October 2001.

38 'Armenia, Greece, Iran Pledge Cooperation on Gas Pipeline Project', *Gas and Oil*, 13 March 2001. http://www.gasandoil.com/news/middle_east/0f9eae8d1bdef19d3aa6ffc868c93371. Accessed on 30 April 2002.

39 *Agence France-Presse*, 12 December 2001.

40 *IRNA*, 20 February 2002.

41 Vladimir Socor, 'Iran-Armenia Gas Pipeline: Far More than Meets the Eyes', *Eurasia Daily Monitor*, Vol. 4, no. 56, 21 March 2007.

42 *BBC*, 'Iran-Armenia Open Gas Pipeline', 19 March 2007.

43 Dilip Hiro, *Blood of the Earth: The Battle for the World's Vanishing Oil Resources* (Penguin Books, New Delhi, 2008), p. 247.

44 Ceragioli and Martellini, n. 7.

45 Turkey held Iran responsible for the delays as it said that the metering station in Iran was not ready for operation. *AFP*, 11 December 2001.

46 Thomas R. Stauffer, 'Caspian Fantasy: The Economics of Political Pipelines', *Brown Journal of World Affairs*, Vol. 7, no. 2, Summer–Fall 2000, pp. 63–78. He holds that since there was no link and no mechanism for physical displacement, the claim that it was Turkmen gas passing through Iran was without any foundation.

47 There were insinuations that Russia could revitalise the PKK unless Turkey agreed to the Blue Stream. Rasizade, n. 30, pp. 22–45.

48 Quoted in Michael Lelyveld, 'Caspian: Setbacks in Gas Exports Could Affect Oil Pipeline', *RFE/RL*, 20 September 2002.

49 The Qatari moratorium is expected to be lifted in a couple of years.

50 *UPI*, 8 March 2010; *Gulf News*, 12 September 2008.

51 Gulshan Dietl, 'Musandam: Creating a New Region across the Water', in Steffen Wippel, ed., *Regionalising Oman: Political, Economic and Social Dynamics* (Springer, Dordrecht, 2013), p. 284.

52 Dana El Baltaji, 'Oman Fights Saudi Bid for Gulf Hegemony with Iran Pipeline Plan', *Bloomberg*, 21 April 2014.

53 Sebonti Ray Dudwal, 'Need to Revive Iran India Ties', *IDSA Comment*, 30 September 2014. http://idsa.in/idsacomments/Needtore viveIran Indiaenergyties_srdadwal_300914.html. Accessed on 31 January 2015.

54 *Petroleum Economist*, Vol. 80, no. 6, July–August 2013, pp. 50–51.

55 Khaled Abubakr, 'Future Gas: From North, South or East? Africa and Middle East Perspective', *7th European Gas Conference*, Oslo, 4 June 2013. http://www.igu.org/sites/default/files/node-page-field_file/North-African%20perspective.pdf. Accessed on 31 July 2013.

56 The Nabucco pipeline, which envisaged a 3,300-km long pipeline delivering 30 bcm of gas per year, could never take off – not due to financial shortage, but on political grounds. It was designed to circumvent Russia, diversify gas supplies and delivery routes to Europe. In a counter-move, Russia entered into a deal with Turkmenistan and Kazakhstan to route their gas through Russia; Hungary withdrew over its dispute with the EU over the state budget; Gerhard Schroeder, the German chancellor, did not support the project; and Azerbaijan drove the final nail, when it chose to route its gas from the Shah Deniz II fields through the Trans-Adriatic pipeline.

57 There is a view that Bashar al-Assad refused to sign the deal citing Syria's strong energy relations with Russia. William Engdahl, 'Syria Attraction: Russia Moving into Eastern Mediterranean Oil Bonanza', *RT*, 13

January 2014. http://rt.com/op-edge/syria-russia-war-oil-528; Also Nafeez Ahmed, 'Syria Intervention Plan Fueled by Oil Interests Not Chemical Weapon Concern', *The Guardian*, 30 August 2013.

58 In the wake of fresh discovery of energy assets, Bashar al-Assad had pronounced a 'Four Seas Policy' as the cornerstone of Syria's energy strategy that placed the country at the centre of Mediterranean, Caspian, the Black Sea and the Gulf.

59 The opponents of the project called it the 'Islamic Pipeline' in derision.

60 Gulshan Dietl, 'The Great Gas War over Syria', *IDSA Comment*, 9 September 2013. http://www.idsa.in/idsacomments/TheGreatGasameoverSyria_gdietl_090913

61 Mike Whitney, 'Et Tu, Mullah? Did Iran Just Knife Putin in the Back?', *Information Clearing House*, 18 August 2014. http://www.informationclearinghouse.info/article39446.htm; Peter Lvov, 'Qatar's Great Power Games', *Oriental Review*, 30 March 2013. http://orientalreview.org/2013/03/30/qatars-great-power-games/. Accessed on 1 September 2014.

62 The *Voice of Russia* carried it with the title 'Bandar Threatens President Putin with Sochi Terrorist Attack', 29 August 2013. http://voiceofrussia.com/2013_08_29/Bandar-Bush-threatens-President-Putin-with-Sochi-terrorist-attack-2596/?from=menu. Accessed on 21 September 2013.

63 *Fars News Agency.* Quoted in http://pipelinesinternational.com/news/construction_commences_on_iran-iraq-syria_gas_pipeline/078918/. Accessed on 20 December 2014.

64 *Tehran Times*, 21 October 2013.

65 *Fars News Agency*, Quoted in *Iraq Business News*, 9 September 2014. http://www.iraq-businessnews.com/tag/gas. Accessed on 10 December 2014.

66 *The World Bulletin*, 8 May 2014. http://www.worldbulletin.net/news/135757/iran-could-supply-gas-to-europe-if-desired. Accessed on 8 June 2014.

67 Quoted in *Tehran Times*, 15 April 2014.

68 Whitney, n. 61.

69 *Caspian Research*, 11 July 2014. http://caspianresearch.com/2014/11/07/iranian-gas-cant-save-europe/#sthash.TASjzk5Y.dpuf. Accessed on 2 January 2015.

70 For more on the Nabucco pipeline, see Chapter 4.

71 'Nabucco Redux?', *Natural Gas Europe*, 6 October 2014. http://www.naturalgaseurope.com/southern-gas-corridor-iran-nabucco. Accessed on 16 November 2014.

72 'Iran Won't Replace Russia as Top Gas Supplier: Tass Quotes Rouhani', *Reuters*, 4 October 2014.

73 *Sofia News Agency*, 30 April 2015. http://www.novinite.com/articles/168258/Iran+Tabled+Rebirth+of+Nabucco+at+Meeting+with+Bulgaria+PM,+Ambassador+Says#sthash.4TNlHa5N.dpuf. See also 1 May 2015. http://sputniknews.com/business/20150501/1021582834.html. Accessed on 14 May 2015.

74 A subsidiary of AXPO, the EGL was taken over by the AXPO in 2012. It has since been renamed as AXPO.

75 The deal was much against the US wishes. *Press TV*, 26 April 2015. http://www.presstv.com/Detail/2015/04/26/408210/Iran-Bulgaria-

Nabucco-gas-Europe-pipeline-Russia-sanctions-nuclear-nuclear-talks-nuclear-agreement-Mohammad-Javad-Zarif-Zarif-John-Kerry-Kerry-Obama-Rouhani-Senate-Congress-White-House-Tehran-Washington. Accessed on 25 May 2015.

76 'Swiss Adopt EU Sanctions on Tehran', *Jerusalem Post*, 26 May 2015.
77 *Reuters*, n. 72.
78 Joshua Kucera, 'Nabucco Carrot and U.S.-Iran Engagement', 5 May 2009. http://www.eurodialogue.eu/The-Nabucco-Carrot-USA-Iran-Engagement. Accessed on 20 May 2014.
79 Martha Brill Olcott, *The Geopolitics of Natural Gas: Turkmenistan: Real Energy Giant or Eternal Potential?* (Center for Energy Studies, Rice University's Baker Institute and Belfer Center for Science and International Affairs of the Harvard Kennedy School, 10 December 2013), p. 22.

4

RUSSIA

An energy superpower?

Russia plays a pivotal role not just because it holds vast
gas resources but also because it sits astride the large
European and emerging Chinese markets and can also
deliver the LNG to the US.[1]

While Russia will remain important to the economies
of the Central Asian and Caucasian states, all of them
are embarked on a rapid diversification of economic and
trade interests that ultimately will pull them east, west
and south, that is, away from Russia. Russia's influence
already diminished everywhere, will probably continue to
fade. . . . On the other hand, Russia will remain militarily
weak, politically incoherent, economically destitute and
demographically sick.[2]

Russia, with the second-largest stock of natural gas in the world, is poised
at a precarious point in its history. In clinical terms, it is at a crisis from
which a patient either improves or collapses. The previous two quotes
point to an unpredictable scenario. This chapter proposes to look at the
Russian gas pipelines that would largely determine the Russian future in
view of the fact that its economy and its global role would largely depend
on the gas exports that are mainly through the pipelines. Two major
pipelines deserve close scrutiny in this respect: one, the South Stream
Pipeline that has run into obstacles and undergone a change of its name
to Turkish Pipeline, and two, the twin pipelines to China that have still
to be named and finalised after the deal was signed in May 2014 between
Russia and China. The sources, the routes, the destinations and the
prospects of these pipelines are diametrically opposite to each other and
bring out the situation of the Russian gas pipelines in the most holistic
framework.

Gas and the Gazprom

The Russian energy history can be traced back to late 1800s. By the 1950s, energy had become one of the major pillars of its economic and political strength. With 77.4 billion barrels of proven oil reserves and an astounding 44.8 trillion cubic metres (tcm) of natural gas reserves – the largest such reserves worldwide and almost one-quarter of the global total – Russia is the most influential energy producer. The majority of these reserves are located in Siberia, with the Yamburg, Urengoy and Medvezh'ye fields alone accounting for more than 40 per cent of Russia's total reserves, while other significant deposits are located in northern Russia.[3]

Over the course of the past decade, Russia has steadily produced between 500 and 600 billion cubic metres (bcm) of natural gas per year, with production topping out at 601.7 bcm in 2008 and amounting to 588.9 bcm in 2010. All told, in 2010, Russian natural gas production accounted for close to one-fifth of the global total, while maintaining a reserve-to-production ratio of 76, signifying the immense potential for growth in the natural gas sector.[4] Energy export revenue accounts for over 70 per cent of its total export revenue and half of its federal budget revenue.[5]

Russia is Gazprom and Gazprom is Russia, so goes the general perception. The two seem indistinguishable as the latter is seen as a giant in positive terms and a monster in not so positive terms. A state within a state, for some. A battering ram to break down defences against the energy imperialism, that is Russia, for others.[6] Gazprom was born out of the Soviet Ministry of Oil and Gas in 1989. The name Gazprom is a short form of *Gazovaya Promyshlennost* (Gas Industry). Today, it is the largest extractor of natural gas in the world that controls more than 60 per cent of gas reserves and 80 per cent of gas production in Russia. Several large subsidiaries, including gas and power companies are controlled by it. And it even owns several media outlets, including the National Television (NTV), a popular television station. In total, Gazprom's profits constitute about 10 per cent of Russia's GDP. Perhaps that is why the company – which even has its own anthem – is considered a bellwether of Russian power.[7] Together with the oil pipeline company Transneft, Gazprom enjoys the right to employ its own armed operatives that could use weapons similar to those enjoyed by interior ministry security guards.[8]

Over time, it has evolved from a national monopoly to a global corporation diversifying into energy super holding including drilling, transportation, oil-and-gas processing and distribution of the end products to the consumer markets.[9] With the Russian plans to use the largely untapped vast resources in the east, Gazprom was appointed as coordinator of the 'Eastern Gas Programme' in order to develop production, transmission, gas processing and chemical industries in East Siberia and the Far East. The strategy envisaged the creation of production centres in Sakhalin, Yakutia (Chayanda field), the Irkutsk region (Kovykta field) and the Krasnoyarsk region, which would be connected to one centre.[10] Russia's resolve to expand east is not new and Gazprom became the linchpin of the eastward expansion project. That was in 2007.

In 2012, Gazprom was assigned the task of building a 3,200-km pipeline from East Siberia to the Pacific port of Vladivostok in an attempt to reduce the country's reliance on exports to Europe and develop closer ties with the Asian customers. According to the Gazprom Chief Executive Alexei Miller, $24.5 billion would be spent on the pipeline and a further $13.7 to develop the Chayanda gas field.[11]

That year possibly marked the peak of Gazprom power. The next year, President Vladimir Putin signed into law a landmark amendment[12] ending the monopoly of the state-owned Gazprom in the export of liquefied natural gas (LNG). More was to follow. When the mega deal with China was finally signed, Gazprom was not considered up to the task singlehandedly. On 22 July 2014, Putin signed a second historic order that ended the Gazprom monopoly over gas pipelines. The companies like the Rosneft, a fellow state-owned energy giant and Novatek, Russia's largest private gas producer, were authorised to use the Gazprom pipelines in Siberia and the Far East.[13] By that time the two had already broken the Gazprom monopoly over shipping gas aboard by receiving permission to export the LNG, although they did not have access to the pipelines.

The company has steadily been stretching itself beyond its capacity calling into question its obligations to its foreign partners. Its purchase of media sponsorship of Olympic sports came in for strong criticism, as did its involvement in the gasification of Russia's regions.[14] Its inadequate attention to the maintenance of old pipelines, technology and innovation are also criticised.

The overall health of Russian economy, including pervasive corruption and deteriorating infrastructure hinder the Gazprom functioning. There are two additional constraints on its functioning. One, unlike the private producers like Novatek and Itera, Gazprom is subjected to price regulations and cannot charge the free market rate. Two, it is mandated to meet the internal Russian demand even at the expense of defaulting on its commitment to its foreign customers.

Gas policies

On 23 January 2014, the Russian Ministry of Energy published the *Main Statements of the Energy Strategy of Russia* for the period up to 2035. At the international level, the priorities identified are (1) promotion of further development of common energy markets within the Eurasian economic space and common principles of regulation; (2) overcoming the crisis in relations with European consumers of natural gas (such developments as 'declining demand, increased competition, transformation of the pricing model; unfavourable regulation; court trials') by means of adaptation of the contract system to current tendencies with regard to Russia's interests; and (3) accelerated entry to the Asia-Pacific market in line with the product diversification of exports.[15]

The Russian plan to extend its reach eastward is already noted. There are distinct differences in Russia's trade policy with Europe on the one hand and with the Central Asian countries on the other. Russia's basic policy desire in terms of Central Asian gas is to purchase the gas at the Russian border rather than permitting transit through its territory. The proviso here seems to be that if the Central Asian countries are willing to sell gas directly to the Commonwealth of Independent States (CIS) customers; Gazprom will allow transit through the Russian territory. Armenia, Azerbaijan, Belarus, Kazakhstan, Kyrgyzstan, Moldova, Russia, Tajikistan and Uzbekistan are members of the CIS. Turkmenistan and Ukraine are associate members and Georgia has withdrawn from the group. If they want to sell gas outside the CIS using Russian pipelines, Gazprom is unlikely to allow it. This is because, although Russia has vast natural gas reserves and spare production capacity, the cost of gas produced from new frontiers such as the Yamal Peninsula would be higher than that of gas from Central Asia.[16] The Russian gas, in the situation, would lose out to the rival supplies from Central Asia.

For Russia, it makes eminent sense to buy the Central Asian gas so that it can free up its own Sakhalin and Siberian Far-East reserves for supply to Japan and South Korea. In the process, Russia can delay development of more expensive fields such as the onshore and offshore fields of the Yamal Peninsula. Therefore, Russia is keen to regain its sphere of influence in Central Asia and act as the energy leader of the region.[17] The Central Asian states can be the swing producers till the time that the Yamal Peninsula and the offshore Shtokman field in the Barrent Sea are developed.[18] The declining production in the traditional fields like Urengoy and Nadym can be supplemented by an additional source of supply.

Russia plays a pivotal role not just because it holds vast gas resources but also because it sits astride the large European and emerging Chinese markets. Where Russia wins on location, it loses out on technology. The Russian energy experts are aware of the technology deficit in the country. Alexei Mastepanov, deputy director of the Institute of Oil and Gas

Problems, Russian Academy of Sciences, is a vocal advocate of technology in the Russian energy sector; particularly in the energy transportation. He laments:

> If we acquire efficient long-distance transportation technologies for traditional fuels, hydrocarbons will retain their relevance for many decades to come; if people learn anytime soon to use new locally obtained energy sources we can safely forget about the world oil and gas trade.[19]

Alexander Orlov, another Russian scholar on the issue, adds that while Russia was busy building pipelines and a system of oil and gas extraction in complex climatic conditions, Americans achieved a technological break-through and organised consistent and profitable production of shale gas. They ensured their energy security for many years to come and were pre-pared to become one of the players on the gas market. Europeans using the shale technology was even more worrisome, he said, as they could then achieve diversification of the sources of hydrocarbons and considerably trim the role of Russia.[20] The technology deficit, in short, meant a competing producer in the US and a shrinking market in Europe.

The Russian energy policies are interpreted in vastly different ways. Do they prioritise energy as a foreign policy tool, or are they more focused on the revenue generating aspect? In normal circumstances, the two could be contradictory most of the times. The STRATFOR[21] holds that the global and regional circumstances have changed to the point that Moscow has had to prioritise one of the two uses of its energy industry – and it has unequivocally decided to maintain its revenue-generating capability. The Kremlin has begun crafting a set of policies designed to adjust the country to the changes that will come in the next two decades, it says.[22]

There are multiple levels at which the Russian revenue-generating policies can be understood: different prices for different markets, limit-ing resales, take-or-pay provisions[23] and so on. Economics and diplomacy are employed in this pursuit. Two instances bring out a combination of economic incentives and diplomatic finesse that has brought Russia huge rewards in gaining gas assets in the neighbourhood.

In 2011, Gazprom bought the remaining stake in Belarus' gas pipeline system to become its sole owner in a move that strengthened Moscow's control over gas exports to the West. It already had owned 50 per cent in Belarus' pipeline operator, Beltransgaz, and wanted to gain full control. It said in a statement after the signing that it agreed to pay $2.5 billion for the remaining 50 per cent stake. In addition, it would provide Belarus with a $10 billion loan spread over the next decade to help it pay for a nuclear power plant that would be built by Russia. As part of the deal, Russia reduced the price of gas to Belarus to $164 per 1,000 cubic metres of gas in the first quarter of next year, down from the $280 per 1,000 cubic metres it was paying in the third quarter of 2013.[24]

Two years later, Kyrgyzstan approved a deal to sell the country's debt-ridden natural gas monopoly to Gazprom for $1. The decision, backed by 78 deputies in the 120-seat parliament, handed Moscow control over a strategic asset in the Central Asian state in exchange for a guaranteed supply of fuel. Under the agreement, Gazprom would gain control over pipelines, gas distribution stations and underground storage facilities owned by Kyrgyzgaz. It committed to invest 20 billion roubles ($610 million) in modernising the Kyrgyz company's infrastructure over the next five years and take over more than $40 million in debt. The deal gave Gazprom full control over Kyrgyzstan's gas sector. The only proviso in the agreement was that the Russian company could only transport its fuel using pipelines belonging to Kyrgyzgaz.[25] The deal was seen as Kyrgyzstan changing its dependency from one country to dependency on a more benign country as it had been dependent on Uzbekistan for its gas supplies since independence.[26] By taking away one of its steady gas customers, Moscow sent a signal to Tashkent not to follow too independent a foreign policy or nourish dreams of becoming a regional power.

A diametrically opposite view on Russia's gas strategy holds that Russia leverages its energy to achieve its broad strategic goals.[27] Energy is a foreign policy tool as energy security is absolutely vital to its vision of the country's future. The Russian Energy Strategy, adopted in 2003, enshrined this philosophy: The energy strategy aims to maximise the effective use of natural energy resources and the potential of the energy sector to sustain economic growth, improve the quality of life of the population and promote strengthening of foreign economic positions of the country.[28] The Russian aggressive diplomacy vis-à-vis Georgia, Ukraine and its own region of Chechnya adhere to the objective of strengthening its foreign economic position, as spelt out.

The Russia–Georgia war in 2008 can be traced back several years before the actual conflict. There were two cut-offs of gas to Georgia in 2001. Georgia was in the middle of a volatile dispute with the separatist region of Abkhazia and the Georgian president Eduard Shevardnadze had requested the Organisation for Security and Cooperation in Europe (OSCE) envoy to replace Russian peacekeepers with a multilateral force, a move that would have weakened Moscow's presence in the region. The country became the first major theatre of confrontation between the US and Russia and the first country to witness a Colour Revolution. The Rose Revolution overthrew Shevardnadze and brought in Mikheil Saakashvili.

The Itera had already bought 50 per cent of the Georgian gas distributor Gruzgas and was in the process of purchasing the remaining 50 per cent.[29] The events in 2008, therefore, had antecedents in dispute over gas. In the course of the war, several Russian bombs fell very close to the Baku–Tbilisi–Ceyhan (BTC) oil and Baku–Tbilisi–Erzurum (BTE) gas pipelines. Many Western analysts suggested that one reason Russian aviators dropped a bomb only 50 metres from the pipelines was to highlight how insecure that link between the Caspian and the West which bypasses Russian territory had become. The counter-argument held that if Russia wanted to bomb the lines, it would have kept trying until it hit the target rather than giving up after one attempt; also it was not in Russia's interest to bomb the pipelines as doing so would have badly alienated Azerbaijan.[30] Deliberate or not, the pipeline came sharply into focus during the five-day conflict. Without acknowledging that sending such a message was among the Kremlin's war aims, Moscow commentator Aleksandr Shustov argued that 'one of the important consequences of the war' had been growing recognition by all parties of just how 'insecure' all pipelines and other transportation arteries through Georgia were.[31]

Ukraine's location between the major gas producer Russia and the major gas consumer Europe, its large transit network and its large underground storage capacity[32] makes it a fiercely contested space. The gas pipelines through Ukraine have been a bone of contention between Russia and the US since the break-up of the Soviet Union. The Turkmenistan–Russia–Ukraine gas pipeline, in the Soviet era, was a domestic grid. In the late 1980s, the Soviets exported oil and gas through Druzhba (Friendship) pipeline to Eastern Europe via ports on the Black Sea (Novorossiysk, Odessa and Tuapse) and the Baltic (Ventspils). Odessa went to Ukraine and Ventspils to Latvia when they became independent. The gas grid, accordingly, crossed three sovereign states: Ukraine, Russia and Turkmenistan.

In the post-Soviet era, it has become an illustration of the triangular dependencies of the three countries. All of Turkmenistan's gas exports outside Central Asia passed through Russia, which put the latter in near complete control of around three quarters of Turkmenistan's exports. Russia's position was similarly vulnerable vis-à-vis Ukraine in that 90 per cent of its gas exports to Europe passed through that country. During the 1990s, the Russian deliveries to Ukraine were the equivalent of 55–80 per cent of its entire export volume to Europe. Thus Ukraine was the transit point as well as the choke point of the Turkmen and Russian exports. It had also been a leaking point of the deliveries.

In early 1990s, there were serious disruptions as Ukraine pilfered gas for its own domestic use. The problem came to a head as the Turkmen suspended supply over a price dispute and Russia over accumulated debt. As a result, the importers in France, Germany, Italy, Turkey, Romania and Bulgaria were stranded with much lower amounts of deliveries than agreed to. In order to stabilise the situation and avoid similar problems in future, Russia signed an agreement with Ukraine in April 1994 whereby it would deliver 10 bcm of gas to be stored in two Ukrainian facilities for delivery to European customers in the winter months. In addition, one of the major Ukrainian transmission companies stated that it would not interfere with gas in transit to Europe. The gas deliveries became an important issue in the political and security relationship between Russia and Ukraine, having featured in the package of agreements which included issues such as the future of the Black Sea Fleet[33] and Ukrainian nuclear weapons.[34]

In October 2000, the then European Commission President Romano Prodi declared that Europe intended to double the import of energy from Russia over the next 20 years. At that time, the share of Russia in the total consumption in Europe already stood at 26 per cent. The already signed contracts on a long-term basis would ensure that Russia kept this share in the future. The basis of Gazprom's competitiveness on the international market was its gas supply reliability based on resource base, the united transportation system, diversification of export routes and co-operation with partners.[35]

Prodi's announcement swung the concerned countries into action. The result was a document signed by the Russian president Vladimir Putin, the Ukrainian president Leonid Kuchma and the German Chancellor Gerhard Schroeder on 12 June 2002. It envisaged a consortium of Russia's Gazprom, Ukraine's Naftogaz Ukrayiny and Germany's Ruhrgas. The consortium was to attract investment worth $2.5b for renovation of Ukrainian pipeline and $15b for the development of infrastructure network in the 10 years thereafter.[36]

The infrastructure that brought around 40 bcm gas to Russia and Ukraine in the 1990 from Central Asian reserves had remained under-utilised by around 20 bcm per year. The Kazakhstan's Karachaganak gas field lies less than 300 km from the existing Orenburg pipeline which – with the depletion of the Orenburg field – could eventually allow up to 30 bcm of Kazakh gas to be brought to Europe after 2000.

Thus around 50 bcm of Central Asian gas can be brought into Europe with very little more than refurbishment expenditures on existing infrastructure. Central Asian gas could either be delivered to Europe by displacement – which would undoubtedly be the preferred option of sellers in both Turkmenistan and Kazakhstan – or sold in Russia and Ukraine, thereby freeing up more Russian gas for sale in Europe (almost certainly the preferred Russian option). The eventual outcome would have been a compromise between these positions.[37]

Gazprom took additional measures to enhance its reliability and flexibility by diversifying the transport routes and consolidating co-operation with its partners. While the pipelines crossing Ukraine already handled some 80 per cent of its exports and had capacity of around 130–140 bcm a year; the relatively new Yamal–Europe line, across Belarus and Poland could handle about 35 bcm a year.[38]

In August 2002, Russia and Ukraine signed a 30-year gas deal on the lines of the 25-year Russia-Turkmen deal of April. Under its terms, Russia agreed to deliver gas to be transited through a yet to be built Bogorod-chany–Uzhgorod line. Initial deliveries in 2005 were scheduled to be 5 bcm, and were to surge up to 19 bcm in 2010. At the end of 30 years, there was a provision to extend it for five more years. In addition, Russia agreed to operate the Ukrainian gas-transit network; expand the existing gas line linking Novopskov in eastern Ukraine with Uzhgorod in the far west; and build a new line from Bogorodchany to Uzhgorod.

In January 2000, the then US Secretary of State Madeleine Albright identified Ukraine as one of the states deserving Washington's particular notice and assistance, promising to help sustain 'this partner and friend's democratic path'. The US proposed to assist Ukraine from a security standpoint, enabling it to better withstand Russian pressure. More specifically, the US assistance was expected to secure the Ukrainian export route. The US stood to gain from it as Ukrainian route would provide a secure and reliable complement to the Baku–Tbilisi–Ceyhan (BTC) under consideration at the time. It did not require the US to abrogate its commitment to Turkey, but it would nonetheless serve as an excellent hedge should the BTC fail.[39]

The BTC construction began in 2002. It would be one of the great engineering feats. It is about strategic geography of Central Asia rather than a mere pipeline. One thousand kilometres long, the pipeline was expected to carry 1 mbd of oil by 2008. Ironically, the pipeline might not have adequate oil to carry by then. Since there is not enough oil in Azeri sector, the supply from Kazakhstan would have to be brought in to make it viable. The BTC, therefore, proposed to secure the link from the Kazakh port of Aktau, across the sea from Baku, to be linked via underwater pipeline to circumvent Russia. Russia, on its part, insisted on a ban on the pipeline along the bottom of the sea to be included in a future convention on Caspian Sea's legal status – on environmental grounds.[40]

The Center for Strategic and International Studies, a prestigious think tank in the Washington DC, published a report on the Russian energy policies. It warned against the 'neocolonial characteristics of Russia's foreign energy policy' and advocated that the US should take the lead in working with the EU and the Central Europeans to better understand the political and security risks that stem from Russia's use of energy as an instrument of foreign policy.[41] The report served to underline the US stakes in Ukraine. With 1,650 troops in Iraq, Ukraine was the fourth largest contingent in the US war on Iraq at that time. The pro-Russian Ukrainian president Leonid Kuchma and Prime Minister Viktor Yanukovych, however, remained the stumbling blocks to a long-term Western policy towards the country. It was pointedly excluded from both the EU and the NATO expansions and also from the list of possible invitees.

The struggle for influence over Ukraine intensified in mid-2004. In July, the Ukrainian government announced that contrary to an earlier decision made in February to use the $0.5b Odessa-Brody pipeline to transport oil from Kazakhstan and Azerbaijan to markets in Western Europe, it would reverse the flow so that up to 14b tonnes a year of Russian oil can be transported to Odessa, from where it will be shipped through the Bosporus Straits. The US Ambassador in Ukraine reacted sharply to the development by remarking that by reversing the flow of the pipeline Ukraine had missed a golden opportunity to take a step forward in the independence of its energy sector.

Putin paid a visit to the country in the middle of high drama – to exhibit solidarity and to extend support to a beleaguered ally. In a joint press conference with Kuchma in Yalta, he told the West in a manner reminiscent of the Cold War not to get in the way of Russia and Ukraine forging closer ties. 'Foreign forces, both inside and outside the country, are trying everything possible to compromise integration between Russia and Ukraine,' he said darkly.[42] He also tried to influence the presidential elections in Ukraine in favour of the then prime minister Yanukovych.

The 'Orange Revolution'[43] radically changed the situation in Ukraine. It was the second Colour Revolution after Georgia. In late 2004 and early 2005, there were massive protests against the elections that were seen to be rigged in favour of the victory of Yanukovych over his contestant Viktor Yushchenko. The re-vote resulted in favour of the latter. Yanukovych and Yushchenko were the favourite candidates of Russia and the US, respectively. The newly elected president Yushchenko praised the US as bedrock of support for democracy and rule of law and expected it to help Ukraine get into the European Union (EU) and the World Trade Organisation (WTO). And should Russia start flexing its hegemonic muscles, Ukraine would appreciate Washington's backing.[44]

A serious stand-off between the two followed, when the Russians cut off gas supplies to Ukraine over price dispute.[45] A compromise was reached only after Ukraine agreed to pay more for the gas that was, till then, highly subsidised. There was a second serious stand-off in 2009. And a third in 2012, when Ukraine's parliament barred any sale of NAK Naftogaz Ukrainy after officials said Russia sought a stake in the state-run energy company's natural-gas pipelines in exchange for cheaper supplies of the fuel.[46] Within a few months, the crisis deepened. Russia dropped its gas exports as its own domestic consumption soared in an unusually cold winter. Gazprom had an obligation to deliver gas to the Russian consumers on a priority basis. Across Europe, 300 people reportedly died of cold even as the temperature in Ukraine plunged to minus 32 degree Celsius.[47] The German energy giant RWE seized the opportunity and began supplying Russian gas after it had crossed Slovakia through Ukraine back to Ukraine through a reverse pipeline.[48] Since there was no destination clause written into the contract between Gazprom and the RWE that would have prevented the resale, it was a legal deal. The diplomatic environment deteriorated even further thereafter.

The pipeline suffered a major setback once again; this time at the Ukraine end. In 2013, Ukraine imported approximately $10 billion worth of natural gas from Gazprom and owed $3.3 billion at the start of the year. The debt was a significant point of leverage for Russia. In December 2013, the deal reached between Vladimir Putin and Viktor Yanukovych partly around gas debts touched off broader protests. Russia agreed to buy Ukrainian government bonds to pay for the debt and reduce the price of gas by about a third. When Yanukovych left Ukraine, Russia cancelled the remaining bond purchases and announced the gas price would rise beginning 1 April 2014. By November, the government decision to delay an association deal with the EU fuelled more protests across the country. War of words ensued, and went beyond the words.

The Euromaidan protests in the Ukranian capital Kiev resulted in the ousting of Yanukovych in February 2014. Russian forces annexed and incorporated the country's Crimean Peninsula the next month. Crimea received most of its electricity (90 per cent) and other energy supplies (25 per cent of its gas) from Ukraine. This dependence resulted in further Russian actions to secure Crimea's supplies, which had the potential to aggravate tensions or trigger a broader conflict. On 15 March, Russia seized a gas terminal in mainland Ukraine just across the border from Crimea, presumably because of supply security concerns in Crimea.[49]

Guenther Oettinger, the EU's energy commissioner sought to mediate in the dispute, but failed. The entire country is in a fratricidal conflict with no end in sight. There already is a de facto division of the country.[50] The gas trade between the two is still going on, as Russia agreed to extend the supply of discounted gas for three more months on 31 March 2015. Under the January–March 'Winter Package', Ukraine paid $329 per thousand cubic metres that reflected a discount of $100 for the same quantity.[51] The Russian gas supply to Europe through Ukraine continues, even as Russia has announced its intention to suspend the transit after 2019, when the contract expires.[52] Ukrainians are confident of continuing the transit even beyond that.

The search for an alternative route has continued, as well. Ukraine's parliament approved a law in March 2014 to allow gas transit facilities to be leased on a joint venture basis with participation from firms in the European Union or US. Ukraine would hold 51 per cent and foreign partners would be offered 49 per cent in the venture, which would manage both transit pipelines and underground gas storage facilities.[53] In December, the European Bank for Reconstruction and Development extended a loan of 150 million Euros to Ukraine to modernise its section of the pipeline. Almost 120 km of the pipeline would be replaced and two compressor stations would be modernised under the project that would be completed in four years. A further loan of 200 million Euros would be made within the stipulated time frame.[54]

The Russian control to retain its former space and the US diplomacy to extend its control over the same have repeatedly come to collusion. Europe is greatly dependent on Russian gas flowing through the Ukrainian territory and has high stakes in Ukraine as well. In short, the country is and will always remain a fiercely contested space among external actors. A vivid description of the country goes like this:

> While grand strategies are plotted in Washington, Brussels, Moscow and Beijing, victories are won on the ground, using all the weapons of globalisation: money, pipelines, diasporas, and the media. Ukraine is a country where this diplomatic game is being played around the clock by politicians, generals, activists, and business people.[55]

Unlike Georgia and Ukraine, the war in Chechnya was against the separatists within. The Russian–Chechen conflict goes far back in time. In recent years, the Chechen declaration of independence in 1991 brought severe retaliation from Russia three years later. The energy aspect of the war was equally important in the Chechen case. The timing of the Russian action is seen to be linked to the development in Azerbaijan, according to an analysis.[56] It came within a few weeks after Azerbaijan had signed the 'Contract of the Century' with transnational energy corporations, which would have loosened Russian control over Baku from where a pipeline running straight across Chechnya would have brought oil and gas to the Russian port of Novorossiysk on the Black Sea. The Chechen capital Grozny was second only to Baku as the biggest oil town in the Soviet Union. Two years of fighting was brought to an end in 1996 under an accord that preserved Chechnya's de facto independence. In its efforts to establish itself as the

main Caspian export route, Russia sought to obtain the Chechens' cooperation to let the Azeri and Kazakh oil and gas pass through its territory in
return for a share of the transaction. Hence between 1994 and 1999, the
Russian route was either open or closed to Caspian oil depending on the
status of Russian–Chechen relations, including the presence or absence of
military operations in that republic, and of course, technical problems.[57]

Vladimir Putin initiated an offensive within two weeks after becoming
the prime minister in August 1999. He was determined to address the
uncertainty about the availability, reliability and safety of the route. Apart
from their obvious objective of restoring their sovereignty over their breakaway republic, what motivated the Russians to resort to a military solution
in Chechnya one more time was their growing concern about losing the
chance of becoming a major export route because of lack of reliability. This
was especially a concern at the time when Western energy companies also
had political reasons not to opt for the Russian route. The massive military
operation of August 1999 lasted officially until early 2000 when the Russians captured Chechnya's capital Grozny.[58]

In the first war, the oil considerations were obviously present.[59] Accordingly, federal forces made every effort to spare the oil and gas infrastructure
from combat destruction. For the Russians, the pipelines passing through
Chechnya had to be safe; otherwise the Western energy companies would
look for other means to evacuate the Caspian Sea oil and gas. For the
Chechens, the oil pipeline was one of the rare sources of revenue and the
hope for the future reconstruction.[60] That provided for maintaining some
oil production throughout the war and restoring the key pipeline in a matter of months after the withdrawal of Russian troops in October 1996.
In contrast, from the very start of the second war, the oil and gas infrastructure was a priority target for Russian forces, which perhaps explains
Moscow's decision to abandon plans for the strategic Baku–Novorossiysk
pipeline going through Chechnya. A decision had been made to build a
312-km bypass line around Chechnya. The bypass pipeline going through
Dagestan (also a highly unstable region) was being built as a reserve option
rather than as an alternative to Baku–Novorossiysk.

Gas pipelines from Russia

In addition to dominating the upstream, Gazprom dominates Russia's natural gas pipeline system. There are currently nine major pipelines in Russia,
seven of which are export pipelines. Three pipelines – Blue Stream, North
Caucasus and Mozdok–Gazi–Magomed – connect Russia's production
areas to consumers in Turkey and Former Soviet Union (FSU) republics
in the east. The other four carry Russian gas to East and West European

markets via Ukraine and/or Belarus. Together, they have a combined capacity of 4 tcf (0.1132 tcm).[61]

Yamal-Europe carries Russian gas to Poland and Germany via Belarus with a throughput capacity of 1 tcf (0.03 tcm). The currently proposed *Yamal-Europe II* would expand the existing pipeline by 1 tcf (0.03 tcm), though disputes between Poland and Gazprom on routing of the pipeline make the project less likely.

Blue Stream is a 750-mile (1,207 km) long pipeline that connects Izobilnoye in Russia to Samsun, Turkey, via the Black Sea. The pipeline's capacity is approximately 560 bcf (16 bcm).

North Caucasus is a 350-bcf (10 tcm) pipeline that runs to Georgia and Armenia. This pipeline is a frequent target of sabotage in the Northern Caucasus.

Yamburg–Uzhgorod, Orenburg–Uzhgorod, Urengoy–Uzhgorod and *Dolina–Uzhgorod* are four pipelines with throughput capacity between 700 bcf (20 bcm) and 1 tcf (0.03 tcm) that carry Russian gas to Western European countries (mainly Germany, Italy, France) via Ukraine.

Gazi–Magomed–Mozdok pipeline connects southern Russia with Azerbaijan. Initially, this pipeline was used to export Russian gas to Azerbaijan but it has been reversed and now it can ship about 200 bcf (6 bcm) of Azeri gas to Russia. It is approximately 400 miles (6,439 km) long.

Nord Stream Pipeline

After Ukrainian disruptions in 2009, Germany worked with Russia to build Nord Stream, a pipeline that sidesteps Ukraine to bring roughly 50 bcm of Russian gas to West Europe each year. It is a 760-mile (1,223 km) offshore pipeline that runs between Vyborg, Russia and Greifswald, Germany, along the Baltic seabed. It is the longest sub-sea pipeline. Its throughput capacity is 1.9 tcf (0.054 tcm), and it ships gas from Yuzhno–Russkoye field directly to Germany and northern Europe. The pipeline was launched in November 2011. Construction of a second parallel line was completed in April 2012, which added about 900 bcf (25 bcm) of additional capacity. The pipeline is functioning since the end of 2012. Gazprom claims that the line will provide customers in Western Europe with the most reliable gas deliveries. It maintains that the pipeline route was adjusted with due consideration for special areas, such as environmentally sensitive regions, chemical weapons dump sites, military zones, critical navigation routes and other dedicated areas serving business or recreational purposes. Nord Stream is designed so as not to cross the World War II ammunition dump sites.[62]

Since there are no transit countries for Nord Stream, it reduces Russian gas transmission costs and eliminates any possible political risks. Exactly for the same reason, the pipeline has aroused a lot of suspicions and fear among the countries that have been cut out of its transit – Ukraine, Slovakia, Czech Republic, Belarus[63] and Poland. They are not in the loop, and can therefore be easily starved of the gas supply, as Russia leaps directly to European customers. On the other hand, the European Commission has ruled that the Nord Stream must allow other gas suppliers to share capacities of its pipeline on German territory.[64]

Undeterred by a chorus of protests from the EU members, Russia and Germany announced the Nord Stream II project that is also planned to cross the Baltic Sea reaching Germany directly. On 18 June 2015, Gazprom, Germany's E.on and BASF, and Austria's OMV signed a Memorandum of Understanding that will add another 55 bcm to the Russian gas delivery. The pipeline will have a further negative impact on supply security among the EU members. In addition, it will result in a loss of roughly $3b transit fees that Poland, Slovakia and Ukraine receive from Russia annually.[65]

The Italian prime minister Matteo Renzi – sore over the demise of the South Stream that would have benefitted his country – held up approval of sanctions rollover against Russia in the European Commission till there was a discussion on the Nord Stream II.[66] Arseniy Yatsenyuk, the Ukrainian prime minister, urged the EU to block the project on the ground that it was not in the interest of Ukraine or the EU.[67] The Ukrainian President Petro Poroshenko termed the pipeline the country's 'greatest concern'. Amos Hochstein, the US Special Envoy for international energy affairs, perceived 'an overarching political agenda to get rid of Ukraine as a transit country at all costs'.[68] An incensed chairman of Slovak minister of the gas pipeline company called Sigmar Gabriel, the German minister of economy, Joachim von Ribbentrop in an apparent reference to the Nazi Germany's foreign minister who signed the notorious Molotov–Ribbentrop non-aggression pact with the Soviet foreign minister Vyacheslav Molotov that included secret provisions of dividing vast areas of Europe into their respective 'spheres of influence'.[69]

In November, Estonia, Hungary, Latvia, Lithuania, Poland, Romania and Slovakia wrote a letter to the European Commission demanding rigorous application of EU energy rules to the project and asking for a summit level meeting to discuss the issue. They also decided to seek redress in the EU courtrooms, from the EU General Court all the way up to the European Court of Justice.[70] An extended judicial process will delay the implementation of the project.

The German response to the collective cry from the rest of the EU members is one of nonchalance. 'It was normal that the EU leaders have different views on some issues from time to time,' the German chancellor Angela Merkel said. 'I made clear, along with others, that this is a commercial project; there are private investors,' she added discounting any role of the German government into the matter.[71] Germany took a legal position that the EU law does not apply to the undersea pipelines – a deliberate oversight of the fact that the territorial waters come under the exclusive sovereign jurisdiction of the littoral states. The Russian response to the problem was more restrained; it agreed to lower the Gazprom stake from 51 per cent to 50 per cent so as not to violate the EU rule on competition.[72]

An extended judicial process is set to delay the implementation of the project even as it is set to further expose the divisions within the EU.

South stream/Turkish pipeline

The project history started in 2006, when Gazprom and Eni entered into the Strategic Partnership Agreement entitling Gazprom to supply Russian gas directly to the Italian market starting from 2007. Under the Agreement, the existing contracts for Russian gas supplies to Italy were extended to 2035.[73] The pipeline would transport natural gas from Izobilnoye in Russia and would run for 560 miles (900 kilometres) under the Black Sea, achieving a maximum water depth of over 6,500 feet (1981 metres). The second, onshore component was to cross Bulgaria to Greece, Italy and Austria. The South Stream project was to be implemented with a view to diversify the routes of natural gas supply to European consumers. An ambitious project to deliver 63 bcm of gas annually, it was expected to be completed and start its commercial operation by the end of 2015.

In January 2008 the South Stream AG joint project company (JPC) was registered in Switzerland. The company was established by Gazprom and Eni on a parity basis. In 2009, the pipeline was rerouted through the Turkish exclusive economic zone (EEZ) to avoid the Ukrainian EEZ as Russia and Ukraine fell out on the terms of the agreement. On 27 November 2009 Gazprom and Électricité de France (EDF) inked the Memorandum of Understanding on EDF's possible participation in constructing the South Stream gas pipeline offshore section. On 19 June 2010, Gazprom, Eni and EDF signed the trilateral Memorandum providing for specific steps towards the French company's entry in the shareholding structure of South Stream AG. The Memorandum contemplated that the entry would be implemented through Eni's stake reduction in the JPC. On 21 March 2011, Gazprom and the German company Wintershall signed a Memorandum of Understanding under which Wintershall would participate in the offshore section of the project and acquire a 15 per cent stake. The Memorandum stipulated conclusion of long-term contracts for natural gas supply.[74]

The next month, Gazprom entered into a Framework of Agreement with the Austrian Company OMV[75] that set forth conditions and deadlines for the implementation of the project. With this, the OMV's Baumgarten distribution node would become 'Central Europe's major turntable'.[76]

The project took off seriously after that. In March 2014, Saipem, a subsidiary of Eni was awarded the contract for laying the first string and for developing landfall and shallow water sections for all four strings of the pipeline. Four parallel pipelines were to be laid across the Black Sea.[77] The next month, three European countries – Serbia, Bulgaria and Hungary – reconfirmed their willingness in hosting the pipeline through their territories.[78] It is noteworthy that the last two are members of the European Union and were under severe pressure from it not to participate in the project. Bulgaria's stand on the issue was to defy the EU and stay with the project. Its political elites saw the EU behaviour as hypocritical given the fact that some EU countries that were demanding Bulgaria drop the project also received gas shipments from Russia. They insisted on an equal right to secure their own economic interests like others.

Within a month after striking an independent posture, Bulgaria retreated. In August, it suspended work on the line till Russia conformed to the EU law.[79] The other countries involved in the project received a rude shock, as they stood to suffer multibillion dollar losses. The Serbian Foreign Minister Ivika Dacic expressed his deep disappointment at Bulgaria's decision in a meeting with the Russian Foreign Minister Sergey Lavrov. Two weeks later, the Bulgarian prime minister Boyko Borisov recanted and stated that the country was ready to issue all necessary permits for the construction of the pipeline, adding that the EU was supportive of his stand and did not want it to suffer financial consequences for stopping the project.[80]

On 1 December 2014, Putin visited Turkey. After meeting the Turkish prime minister Recep Tayyip Erdogan, he pulled the plug on the project by announcing that the pipeline route would be altered to go through Turkey, instead. At a joint press conference with Erdogan in Ankara, Putin made a detailed response to the changed circumstances. He made three specific points: one, the project was at the stage when the construction in the Black Sea must begin, and since Russia had not received an approval from Bulgaria till then, it made no sense to invest hundreds of millions of dollars and then stop when it reached the Bulgarian waters: two, he taunted Bulgaria of not acting like an independent state and advised it to demand loss of profit from the European Commission that it would have been receiving through gas transit to the tune of 400 million Euros a year; and three, that pulling Bulgaria back was not in the economic interests of Europe and it harmed its relations with Russia.[81] He named the new project Turkish Pipeline Project.[82]

Apart from rerouting and rechristening the pipeline, Putin resolved to take two more decisive steps in Russia's gas trade: to promote markets

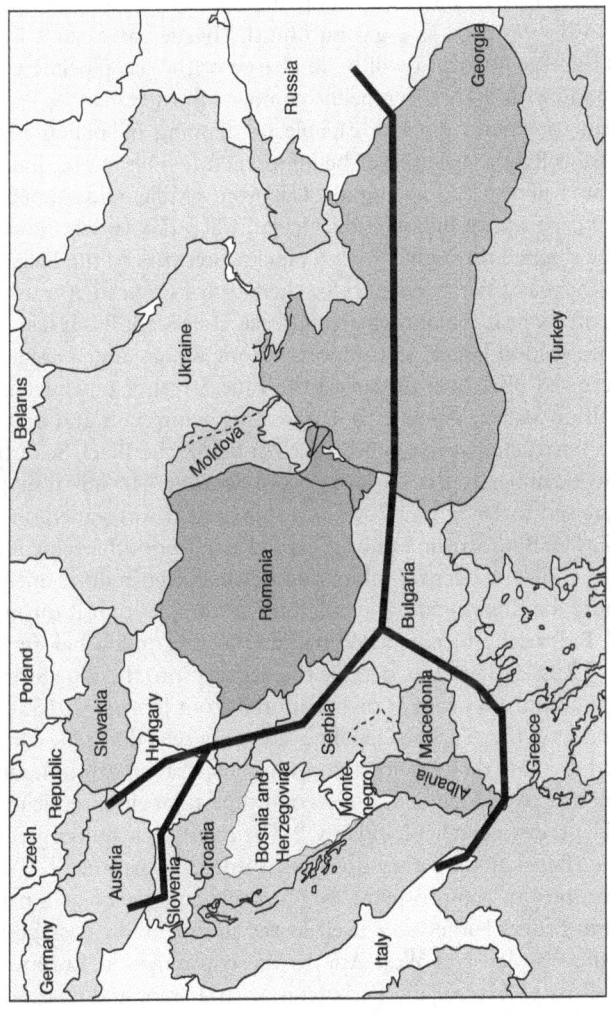

Figure 4.1 Proposed route of South Stream.

Source: Based on http://en.wikipedia.org/wiki/South_Stream#mediaviewer/File:South_Stream_map.png

Disclaimer: Map not to scale.

other than Europe[83] and to speed up projects to export liquefied natural gas.[84]

On that day, Gazprom and Turkish company Botas Petroleum Pipeline Corporation signed the Memorandum of Understanding on constructing an offshore gas pipeline across the Black Sea towards Turkey. The gas pipeline will have a capacity of 63 billion cubic metres, with nearly 50 billion cubic metres to be conveyed to a gas hub on the border between Turkey and Greece. Gazprom Russkaya will be in charge of the gas pipeline construction.[85] Gazprom and Turkish pipeline company Boru Hatlari Ile Petrol Tasima AS plan to build a pipeline capable of shipping 63 billion cubic metres a year from Russia to Turkey. The infrastructure built in preparation for South Stream will be used for this pipe, Gazprom's Miller said. About 20 per cent of that capacity, 14 billion cubic metres, will go to Turkey, he said. The rest will be shipped through Turkey's pipeline network to the Balkans. Russia can build an additional gas hub for the South European consumers on Turkish territory, near the border with Greece, if needed, Putin said.

Turkey is the second-largest gas importer from Russia after Germany. In 2014, it received 27.4 bcm of gas via the Blue Stream[86] pipeline. The 446-km-long Blue Stream pipeline to Turkey was being built at a cost of $3.2b. It was constructed at the depth of 2 km below the Black Sea from Russian port of Dzhubga to the Turkish port of Samsun. The initial supply of 3 bcm was raised to 16 bcm in 2008. The agreement was signed on 22 October 2002. The Blue Stream can be regarded as a major achievement on several counts; it was the deepest submarine pipeline ever built; it opened the second route for Russian gas; it avoided the traditional transit through Romania and Bulgaria; and it reached Turkish market much before the BTC pipeline.[87] Russia offered to increase the delivery into the Blue Stream by 3 bcm per year and a 6 per cent discount on it from January 2015.

The prospects of the Turkish Pipeline are intensely debated. Originally conceived as a line that would circumvent a persistently troublesome Ukraine, it has had to change course a second time to avoid a recalcitrant Bulgaria. Will Turkey be a reliable 'Plan C'? The question acquires greater meaning in the aftermath of Turkey downing a Russian warplane.

A critical scrutiny of the prospects for Turkey's emergence as a natural gas hub is in order. Turkey sees itself as the indispensable passage for energy from the Caspian and West Asia to the consumers in Europe. It imports from Russia, Azerbaijan, Turkmenistan and Iran, and re-exports some of it to Europe. It holds a privileged location between the substantial gas reserves of Russia, Iran, Qatar and the Caspian and the substantial gas markets in Europe.

Its domestic demand is fluctuating impacting its re-export commitment to Europe. The gas infrastructure has been vulnerable to frequent terrorist attacks leading to supply disruptions. The Baku–Tbilisi–Erzerum (BTE) pipeline or any other pipeline entering from Georgia has to pass through Turkey's Kurdish region, where the attacks from the militant rebels have increased in frequency and damage. The pipeline from Iran has also witnessed sabotages.

Regionally, the Interconnector Turkey–Greece–Italy (ITGI) transportation project from the Shah Deniz gas field in Azerbaijan to Italy has had positive fallout in Turkey's relations with Greece. The initial agreement between the Turkish state pipeline company BOTAS and the Greek gas company DEPA led to the laying of the pipeline by the Turkish prime minister Recep Tayyip Erdogan and the Greek prime minister Kostas Karamanlis in 2005 – a milestone in decades of distrust and hostility between the two nations. The pipeline is operational since 2007.[88]

On the other hand, Turkey is entering into troubled waters in its quest for more pipeline options in the region. The BOTAS has already started building a large pipeline towards northern Iraq planning to import gas from the Iraqi Kurdistan. The project is bound to invoke opposition from the central Iraqi government in Baghdad. The Turkish minister of energy and natural resources Taner Yildaz has also expressed interest in playing a large role in importing and re-exporting Israel's gas deposits from the Leviathan gas field. Implementation of the project will have negative fallouts for Turkey's regional policies.

The Qatar–Turkey agreement to build a pipeline originating in Qatar and moving through Saudi Arabia, Jordan and Syria reaching Turkey has been perceived to be one of the major reasons for the uprising in Syria. A pliable regime in Syria would have been an absolute prerequisite for its implementation. The Iraq–Turkey Pipeline and the Arab Pipeline from Egypt have not been realised due to the political instability in the region. The gas imports from Iran have remained static in view of the sanctions on the country.

Beyond the region, Turkey suffers at times and benefits at others from the global scramble for a share in the gas resources. It has been a contested territory for rival pipeline proposals – the most interesting being the erstwhile South Stream project of Russia and the Nabucco project of the West.

The Nabucco, originally proposed in 2002, was to follow roughly the same route as the South Stream. A decade later, it was revised and renamed as Nabucco-West which would have been a shorter and a more modest version. The Nabucco-West would have been a commercially viable and logistically advantageous proposal but the risks of causing a gas glut in the market and of raising tensions with Russia doomed its prospects. In June 2013, its fate was finally sealed when the Trans-Adriatic Pipeline (TAP) was chosen as a route from the Shah Deniz II fields in Azerbaijan to

Western Europe. The TAP will connect with the Trans-Anatolian Pipeline (TANAP) near the Turkey–Greece border in Kipoi, cross Greece, Albania and the Adriatic Sea before coming ashore in southern Italy.

Rotterdam is the undisputed leader as the hub for crude oil. It is one of the world's largest ports with a huge rail, road, air and sea network. Handling the cargo, managing trans-shipment, operating refineries and petrochemical plants, the city has acquired the title of 'Gateway to Europe'. So will Turkey be the hub for gas tomorrow like Rotterdam is the hub for oil today?[89]

Prospects after Turkey shoots down Russian warplane

On 24 November 2015, Turkey shot down a Russian warplane accusing it of violating its airspace. Putin and Erdogan made incendiary statements against each other, which have become increasingly shriller. Incredulously, it does not seem to have any negative impact on the Turkish pipeline beyond a temporary suspension. The very next day, the Russian deputy energy minister Anatoly Yanovsky assured that the gas supplies to Turkey would continue in line with the contract.[90] Within a week after the incident, the Russian minister of economy Alexei Ulyukayev made a surprisingly conciliatory remark on the future of the Turkish Stream and the proposed Russian nuclear power plant in Turkey: 'There have been no decisions at this stage on suspending, freezing or ending financing for these projects. We are working on the assumption that they will be carried out as they were agreed.'[91]

Two days later, the Russian minister of energy Alexander Novak said that work on formulating agreements on the Turkish Stream is 'suspended', because an intergovernmental commission on trade and economic cooperation has stopped meeting under Russia's retaliatory measure against Ankara. He added, as if to smoothen the blow, that the talks on building a nuclear power plant in Turkey remained open.[92] On the same day Alexei Miller, the head of the Gazprom said, 'If Turkey considers it needs this project, it can contact us.' His were the harshest words, in the circumstances. Erdogan adopted a more belligerent posture on the matter. It was Turkey that had suspended the project, he declared, and long before the shooting incident, because of 'Russia's non-compliance with our demand'.[93]

On 17 December, Putin gave his annual press conference to the Russian and foreign reporters. As expected, there was a question on the Turkish pipeline. Putin's response was astounding: we need guarantees in writing from the European Commission that it supports the route. 'If Gazprom's Turkish partners bring us a document of this sort, we can move on.' Not a word about the downed Russian warplane, none about the pipeline being under a state of suspension![94] The Russian need for gas passage outweighs bilateral frictions.

At present, the relations are continuing on a downward trajectory. The resumption of work on the pipeline will remain a hostage to an entire gamut of issues between them, the situation in the region and the EU nod.

The proposed pipelines to China

The Russia–China deal on the gas supply from Russia can be traced back to the agreement of Strategic Cooperation signed in October 2004 during Putin's visit to China. It covered a wide spectrum of joint businesses including discussion of issues related to the gas deliveries to China, implementing joint gas processing and gas chemical projects in eastern Russia and in third countries.[95] In 2007, the Russian Ministry of Industry and Energy approved the 'Development Programme' for an integrated gas production, transportation and supply system in eastern Siberia and the Far East, taking into account potential gas exports to China and other Asia-Pacific countries. In that year, Gazprom broke ground for a pipeline stretching across Russia's Far East and after extension to China, to deliver gas to the country's populous north near Beijing. The pipeline was named the 'Power of Siberia', which could deliver gas via Vladivostok, Russia's eastern port city or via Blagoveshchenk in the Amur region.

In 2009, the two signed a deal under which Gazprom would supply China with 30 bcm gas per year by 2015; although the route was changed to the Altai pipeline and the destination was changed to China's western Xinjiang Province. The Altai route was shorter and more economical. It was a preferred route for the Chinese. The project did not succeed due to price disagreement and competition from other suppliers to China. In addition, the Russians feared that pipeline infrastructure and highway might pave the way for a Chinese expansion into the Altai region. Russia stopped work on this project in 2013 and prioritised the 'Power of Siberia' line.[96]

During President Xi Jinping's visit to Russia in March 2013, Gazprom and China National Petroleum Corporation (CNPC) signed an MoU about delivery of 38 bcm to be delivered from eastern Siberia. The final deal would come later. In the meanwhile, Russia allowed the CNPC to acquire a 20 per cent interest in Novatek's Yamal LNG project in the arctic region, in return for which China agreed to purchase at least 3 million tonnes of LNG from the project. The deal was the first of its kind, as Russia had been reluctant to permit Chinese companies from taking direct stakes in its energy fields.[97]

After long years of negotiations over price, the gas pipeline route and possible Chinese stakes in the project, Russia and China finally crossed the

Rubicon, when Putin and his counterpart Xi Jinping announced a $400 billion natural gas deal in Shanghai 21 May 2014. Gazprom's head Miller signed the deal with Zhou Jiping, chairman of China National Petroleum Corp. Gazprom will provide a trillion cubic metres in gas to the world's second-largest economy over 30 years. This will be the first gas pipeline between the two countries that only traded in the LNG. The Russian LNG accounted for as little as 1 per cent of China's total gas imports till then.

Putin called the deal a 'watershed event' and said implementation would start 'tomorrow'. The final price of the Russian gas was not disclosed, but the informed estimates put it around $350 per thousand cubic metres. In 2013, the average price of Gazprom's gas in Europe was about $380 per thousand cubic metres. The Royal Bank of Canada (RBC) Capital Markets analysts said that the implied terms will give China a steady supply of piped-in Russian gas at a price between 25 and 40 per cent lower than the current cost of importing liquefied natural gas from overseas.[98]

China may contribute as much as $25 billion in advance payments under the contract to invest in the necessary infrastructure, according to the Russian energy minister Novak. Russia will invest $55 billion in the pipeline and the Siberian fields to feed it, while China, responsible for a pipeline on its territory, will spend at least $20 billion, Putin said. 'This is the largest ever contract for Gazprom,' Miller said, adding that the supplies will start in four to six years.[99]

Kremlin lauded the deal as another step in enhancing the 'trust-based dialogue and bringing strategic partnership to new heights'. The contract will be served by Kovykta and Chayanda deposits that hold a total of 3 tcm gas, which is expected to last 30 years, even 50. The gas will be delivered to the eastern part of China, and in the process will make gas infrastructure development across Russia's Far East and eastern Siberia economically viable, the Kremlin site stated.[100]

The next step would be to enter into a similar deal for the western route. The deposits for this route are located in western Siberia. This project is expected to be cheaper. Since all the main issues regarding price calculation, price formulation and government support and benefits for the project have already been addressed by both sides, Kremlin hoped that it could be implemented even faster than the eastern route, if 'greenlighted by China'. The Kremlin compared it with the 'decision of the Soviet Leaders and the Federal Republic of Germany to enter into the famous "gas for pipes" contract'.[101]

The gas for pipes contract refers to the US opposition to Russia's megaproject in the late 1970s to build four 56-inch gas pipelines from Siberia towards the west, connecting into Europe. The US tried to stop the project, blocking sale of US-designed or US-licensed compressors. The

Russians counter-attacked in the early stage of this pipeline war and prevailed. They leaked to the Europeans that the US was conducting sidebar discussions with Russia, offering to waive US opposition to the project if Russia agreed to increase the number of exit visas for Soviet Jews.[102]

It was truly a mega deal, and evoked huge interest. As Russians brushed up the memory of 'Gas for Pipes' during the Cold War, the critics of the deal evoked an even earlier episode. At the other end of the spectrum, it was called 'this century's version of the infamous Molotov-Ribbentrop Pact', a non-aggression pact signed by the then Soviet foreign minister Vyacheslav Molotov and the then German foreign minister Joachim von Ribbentrop in August 1939 that contained a secret protocol that divided vast territories between their borders into their respective 'spheres of influence'. The gas deal was suspected of carving out similar spheres of influence by the two. The pact is already mentioned earlier in the chapter, and shows repeated instances of intense resentment towards German-Russian/Chinese-Russian cooperation.

A comparison was also drawn with the US opening to China in early seventies. 'Putin to Shanghai reprises Nixon to China,' it said. 'Their 1972 strategic coup fundamentally turned the geopolitical tables on Moscow. Putin has now turned the same tables on us.' The Russia–China rapprochement was seen to be undoing the Kissinger–Nixon achievement.[103] It was the US that reached out to China earlier; now it was seen to be Russia doing the same.

The Russians and the Chinese were challenging the post–Cold War international system led by the US, according to this understanding of the deal.

> They are, in deeply troubling ways, changing the world, shaking the foundations of an order that has ensured the absence of major conflict and paved the way for unprecedented prosperity. Their growing cooperation is bound to aggravate their already tense relationships with the United States, and as each of them perceive their future tied to the other, they are bound to see less reason to cooperate with others.[104]

The above was feared in greatly exaggerated terms. Describing the post–Cold War order in terms of absence of major conflict and unprecedented prosperity was a creative interpretation of history indeed.

The overblown fear psychosis contained an element of truth, nevertheless: the gas pipeline deal was not just about the gas. The motives and drivers of each of the two should provide a context for a better understanding. Putin and Jinping could trumpet the deal as their personal accomplishments – to boost their legitimacy at home and to silence the critics abroad. The two could utilise the moment to clip the wings of their giant energy behemoths – Gazprom and CNPC. It was expected that they would likely shuffle leaders and restructure Gazprom and CNPC management as they see fit in order to make this deal strengthen their reputations as firm, decisive national leaders.[105] Speculation was rife around future courses of options before the leaders of the two countries: will Russia continue its policies to encourage Novatek and Rosneft to export LNG to Asia, or will it double down and try harder to protect Gazprom? Will China go ahead with gas pricing reforms that pass on international gas market prices to Chinese enterprises and consumers, who are accustomed to price controls that benefit them, or will it continue to make CNPC and the other energy companies foot the bill for cheap domestic gas? Now that China has one more major pipeline source of natural gas, will it allow CNPC, China Petroleum and Chemical Corporation (Sinopec) and China National Offshore Oil Corporation (CNOOC) to continue their aggressive search abroad for sources of LNG?

The most important question mark hanging on the horizon was: if North China gets most of its gas from Russia and Central Asia, will military leaders in Beijing come to think that the national economy is less dependent on LNG coming through the South China Sea, and thus they can be more bold and adventurous in that contentious arena?[106] The Chinese military officials were thought to be chiming into the debate in Mahanian terms stipulating that 'he who controls the sea controls the world' and that 'it is extremely risky for a major power such as China to become overly dependent on foreign imports without adequate protection'.[107] As the Chinese energy imports rise steadily, defending the sea lanes from the Gulf and the Caspian could become a major security imperative. Would China adopt a more aggressive posture in the South China Sea?

What followed seemed to fuel these apprehensions further. In October, a Chinese delegation led by Premier Li Keqiang signed a package of deals in Moscow in areas including energy and finance. Among the accords was a three-year 150 billion yuan local-currency swap deal, a double-tax treaty, satellite-navigation, high-speed rail cooperation and an agreement on implementing the natural gas contract. Vasily Kashin, a China expert at the Centre of Analysis of Strategies and Technologies in Moscow was quoted to say that Russia was preparing to sign contracts for the delivery of S-400 missile systems and Su-35 fighter jets early next year and may also supply China with its newest submarine, the Amur 1650, and components for products such as nuclear-powered satellites. Since then, the Russian defence ministry has assured that China will receive the first batch of S-400 Triumph long-range anti-aircraft missile systems within the next 12 to 18 months.[108]

Feodor Lukyanov, head of the Moscow-based Council on Foreign and Defence Policy and an Advisor to the Government added a sombre note to the discussion: Russia has long been reluctant to further empower a neighbour that already has four times the economic output and almost 10 times the population. The agreements changed all that and Putin now risks playing a role he's not used to playing: junior partner. 'It's one thing to shift to China when it's just one of several options, but quite another when you have to rely on the Chinese for political reasons,' Lukyanov said. His prognosis was pessimistic: 'There's a risk that Russia will end up considerably under China's sway.'

Omar Lamrani, a military analyst at Stratfor, predicted a conventional arms race in East Asia. 'Japan, Taiwan, the Philippines and Vietnam are already worried about the Chinese developing their military and those concerns will only increase if China gets this Russian equipment,' he said. The S-400, which only Russia currently uses, would extend China's reach to encompass all of Taiwan's airspace, while the Su-35 would allow the Chinese to use the technology to expand their air force, according to his assessment of the weapons to be supplied by Russia.[109]

A second deal followed the mega deal of May. Beyond the substance, the timing of the second deal was startling. Putin and Jinping agreed to a second deal to supply the Russian gas on 9 November 2014. The next day, the US president Barak Obama was in Beijing to participate in the Asia-Pacific Economic Cooperation (APEC) summit. Miller and Jiping signed the Framework Agreement. An initial map released by the Gazprom identified the western route from western Siberia through the Altai Mountains to China's western region of Xinjiang.[110] What was also noteworthy was the fact that the Russian source of gas would be the same that used to deliver gas to Europe before the sanctions came into effect; symbolically shutting the door to the European customers and replacing them with the Chinese. The volume of gas to be delivered would be slightly smaller, that is 30 bcm as compared to 38 bcm via the eastern route. With the two gas deals in place, China becomes Russia's top gas customer and Russia opens up a new market as it risks losing its European customers. The western route was to be accorded a priority in the bilateral gas cooperation, according to Miller.[111]

The Russian end of the deal has economic and also political motives. The draft energy strategy released in January 2014 sets the speed up entry into Asia-Pacific markets as its primary task to the tune of 23 per cent of all its energy exports by 2035. Russia currently ships around 6 per cent of the gas to Asia-Pacific in the form of LNG from its Sakhalin 2 LNG plant, which has a capacity of 10 million tonnes per year. It aims to send 31 per cent of gas by 2035.[112] The gas deal with China assures a timely access to the growing Asian market. The supply to China is slated to begin in four to six years, by which time the global gas market will have substantial levels of supplies from sources like the US, Canada, Australia and East Africa.[113] The Russian gas would not have to compete with the newcomers for customers.

In early 2014, sanctions were imposed on Russia in response to the situation in Ukraine and the Russia annexation of Crimea.[114] In August 2014, Moscow reciprocated by imposing a one-year ban on meat, dairy, seafood and other imports from the US, the European Union and several other countries in response to their sanctions targeting Russia's banking, energy and military industries. The focus away from the West and towards the East was a predictable Russian reaction. The deal with China was its early manifestation; it was also a show of strength.[115]

Dmitri Trenin, the director of the Moscow Carnegie Center, explained the move in a broad perspective: what was originally a 'marriage of convenience' with Beijing has turned into a much closer partnership that includes cooperation on energy trade, infrastructure development and defence. Putin's vision of 'Greater Europe' from Lisbon to Vladivostok, made up of the European Union and the Russian-led Eurasian Economic Union,[116] is being replaced by a 'Greater Asia' from Shanghai to St Petersburg.[117]

The political component is not prominent on the Chinese side; it is a continuation of its pursuit of energy. A net exporter of oil till 1993, China became a net oil importer that year. That makes it one of the late arrivals in the global energy game. Since then its accelerating industrialisation has made it increasingly dependent on imported energy. In May 1997, the then Premier Li Peng advocated Chinese involvement in the exploration and development of foreign energy sources with a view to securing stable, long-term supplies. Since then, China has made investments and acquired exploration and extraction rights in the energy-related fields in Argentina, Bangladesh, Canada, Columbia, Ecuador, Indonesia, Kazakhstan, Malaysia, Myanmar, Mexico, Mongolia, Nigeria, Pakistan, Papua New Guinea, Peru, Russia, Thailand, Turkmenistan, Venezuela and the US. And of course, various countries in the Gulf.

Although China's industrial and commercial growth is in the east and south of the country, the onshore reserves of the coal, oil and hydropower are located in the north and west. As a result, it faces a future of energy supply deficit. By the year 2020, it is expected to be the second-largest consumer of energy in the world – after the US.[118] During the course of the next 25 years, China's gas consumption is expected to grow five times. Though it is self-sufficient in natural gas at present, an active promotion of gas utilisation will push the country into an import dependency. It has consistently advocated a pan-Asian continental land bridge of oil and gas pipelines stretching from West Asia to Southwest and Central Asia, Russia,

Southeast Asia to China. China will be able to cover half of its imports in 2030 from already approved purchase contracts and transmission pipelines from the neighbouring countries. This will help contribute to a relatively high degree of supply security if the bilateral relationships to the export countries remain stable.[119]

The Caspian gas supplies need to be taken on board to put the Russia-China deal in a proper framework. China benefits from its advantage in terms of proximity to the Caspian gas resulting in lower investment and lower expenses because of shorter distance. Russian advantage lies in the existence of Soviet era infrastructure that can be refurbished and utilised for an onward gas supply to Europe. It can also avail of the option to deliver the Caspian gas to Europe by displacement.

Among the Caspian suppliers, Turkmenistan is of special significance in view of its vast resources. The Turkmen gas that goes to China reduces the amount of that much gas that Turkmenistan can load for the onward transfer to Europe. Similarly, more the Turkmen gas for China, less the Russian gas exports to China. In addition, more diversified the Turkmen gas exports and Chinese gas imports, better are their bargaining positions vis-à-vis Russia. Between the options of Turkmen gas going to Europe or to China, Russia would probably prefer that Turkmen gas does not cut into its own near-monopoly of the European market; Turkmen gas to China is a lesser evil of the two.

And lastly, what are the likely medium and long term consequences of the gas deal on Russia itself? China and Russia have been neighbours and have shared a long, chequered history. In late 1600s, Chinese forces drove Russian settlers out of the Amur River valley. Although the border dispute is settled, there still are lingering suspicions of Chinese encroachments into Russia's east. In 2003 Moscow shied away from a Chinese proposal to build a pipeline from Siberia's oil and natural gas fields directly into China. Russian leaders likely believed that it was far better for Russia to build a spur pipeline into China from its trunk line. That way, if relations between the two countries turn chilly again, Russia could still use its trunk line to export its energy resources to other Asian customers.[120]

The tussle over the gas pipeline route – whether eastern 'Power of Siberia' or the western via Altai – is more than simply a matter of economy. A Gazprom analysis states that Russia intends to develop cooperation with Japan and the Republic of Korea in order to avoid full dependence on China.[121] These countries are already buying the LNG from the Sakhalin II project; and there are plans to commission an LNG plant near Vladivostok in 2018. The Russian anxiety vis-à-vis China has a catch–all name: Yellow Peril. It includes the gnawing worries regarding potential mass migration by the Chinese into the vast expanses of Siberia and the Russian Far East and replacing the region with their way of life and values; worries that a stronger China may someday reopen the closed chapter of the border settlement; that Russia may end up as a mere appendage to the Chinese economy by supplying it with raw materials and subsisting out of the income thereof; or, that Russia would incrementally become a 'junior partner' to China in a strategic sense.[122] The prognosis goes even further. Within two decades or less, much of Russian Far East and eastern Siberia might also have been absorbed by China on the strength of Japanese investment in the Siberian gas fields and against minimal Russian resistance.[123]

The Russians have either accepted the inevitable or seem to be confident of managing the Yellow Peril. In January 2015, it announced a plan to give every resident of the country a free hectare of land for 'agriculture, business, development, forestry or hunting' in the vast tracts of sparsely populated Far Eastern region. The land must not be resold to foreign citizens. In a perfectly coordinated move, the authorities in the Zabaykalsky Krai in Siberia, which shares nearly 1,000-km border with China, signed an agreement with the Chinese company Huae Sinban to lease 1,115,000 hectares of Russian farmland to China for a 49-year period to develop agriculture.[124] Most likely, there was an oral understanding on this matter during the gas pipeline agreement itself.

Recapitulation and leads

Russia, with the second-largest stock of natural gas in the world, plays a pivotal role not just because it holds vast gas resources but also because it sits astride the large European and emerging Chinese markets. Gazprom, the Russian energy behemoth is the largest extractor of natural gas in the world that controls more than 60 per cent of gas reserves and 80 per cent of gas production in Russia.

The Russian energy policies are interpreted in vastly different ways. Do they prioritise energy as a foreign policy tool, or are they more focused on the revenue generating aspect? In normal circumstances, the two could be contradictory most of the times. Its revenue-generating policies can be understood at multiple levels: different prices for different markets, limiting resales, take-or-pay provisions and buying stakes or acquiring full control over gas assets in neighbouring countries. Its deals with Belarus and Kyrgyzstan illustrate the policy. The Russian aggressive diplomacy vis-à-vis Georgia, Ukraine and its own region of Chechnya, on the other hand, are examples of Russia leveraging its energy to achieve its broad strategic goals.

The Turkmenistan–Russia–Ukraine is very much more than a gas pipeline and deserves a stand-alone scrutiny. It has been taken up in the next chapter on Turkmenistan, where it originates.

Two major pipelines from Russia deserve close scrutiny. The South Stream Pipeline ran into obstacles when, under the EU pressure, Bulgaria retracted after confirming its willingness to host the pipeline on its territory. In the circumstances, Russia has rerouted the line through Turkey and rechristened it Turkish Pipeline. Originally conceived as a line that would circumvent a persistently troublesome Ukraine, it has had to change course a second time to avoid a recalcitrant Bulgaria. Will Turkey be a reliable 'Plan C'? Even after the downing of a Russian warplane by Turkey, the pipeline project is only suspended, and may be taken up once again when the tensions subside.

Turkey has its own ambition to become a natural hub for the global gas system. It imports gas from Russia, Azerbaijan, Turkmenistan and Iran and re-exports some of it to Europe. It holds a privileged location between the substantial gas reserves of Russia, Iran, Qatar and the Caspian and the substantial gas markets in Europe. Will Turkey succeed to emerge as the gas hub of tomorrow like the Rotterdam is the oil hub of today?

In May and November 2013, Russia signed two mega deals with China in quick succession. The first deal envisages gas delivery from Vladivostok to the eastern part of China through the 'Power of Siberia' pipeline. The second deal will take a western route from western Siberia through the Altai Mountains to China's western region of Xinjiang.

The sources, the routes, the destinations and the prospects of the pipelines to Turkey and to China are diametrically opposite to each other and bring out the situation of the Russian gas pipelines in the most holistic framework.

The pipelines to China have generated tremendous excitement and anxieties. They are perceived as this century's 'Molotov–Ribbentrop Pact' that would shake the foundations of the world order. The Russian themselves regard the deals as moving towards the fulfilment of their vision of 'Greater Asia' from Shanghai to St Petersburg.

Notes

1 Joe Barnes, Mark H. Hayes, Amy M. Jaffe and David G. Victor, 'Introduction to the Study', in David G. Victor, Amy M. Jaffe and Mark H. Hayes, eds., *Natural Gas and Geopolitics: From 1970 to 2040* (Cambridge University Press, Cambridge, 2006), p. 16.
2 Juli A. MacDonald and Enders Wimbush, 'India's Energy Security', *Strategic Analysis*, Vol. 23, no. 5, August 1999, pp. 826–830.
3 *Energy Information Administration*. http://www.eia.gov/countries/cab.cfm?fips=RS. Accessed on 25 February 2015. The British Petroleum estimates are 31.3 tcm, as cited in Chapter 1.
4 Richard Weitz, 'Strategic Posture Review: Russia', *World Politics Review*, 13 May 2014. http://www.worldpoliticsreview.com/articles/13777/strategic-posture-review-russia. Accessed on 15 June 2015.

5 Jane Nakano and Edward C. Chao, 'Russia-China Natural Gas Pipeline Agreement', *Pipeline and Gas Journal*, Vol. 241, no. 8, August 2014. http://pipelin eandgasjournal.com/russia-china-natural-gas-pipeline-agreement? page=3

6 Evan Nate and Economides Michael J., 'Gazprom and Russia's Energy Imperialism', *World Politics Review*, 23 July 2008. http://www.world politicsreview.com/articles/2469/gazprom-and-russias-energy-imperialism

7 Ahmed Mehdi, 'Putin's Gazprom Problem: How the Kremlin Accidentally Liberalised Russia's Natural Gas Market', *Foreign Affairs*, 6 May 2012. https://www.foreignaffairs.com/articles/russian-federation/2012–05–06/putins-gazprom-problem. Accessed on 12 June 2015.

8 'Russia Gives Gazprom Right to Form Armed Units', *Gas and Oil*, 5 July 2007. http://www.gasandoil.com/news/2007/07/ntr73033. Accessed on 23 December 2012.

9 Tatiana Mitrova and Iakov Pappe, 'Gazprom: From the Big Pipe to the Big Business', *The Power of Oil and Gas* (Carnegie Moscow Centre), Vol. 10, nos 2–3, March–June 2006. http://carnegie.ru/proetcontra/?fa=24830. Accessed on 23 November 2013.

10 'Medium Term Oil and Gas Markets, 2011', *Energy Information Administration*, p. 229.

11 'Economy Watch News Desk Team: Russia to Build $38 Billion Pipeline to Asia', *Economy Watch*, 29 October 2012. http://www.economywatch. com/in-the-news/russia-to-build-38-billion-gas-pipeline-to-asia.30–10. html. Accessed on 13 October 2014.

12 'Law Ending Gazprom's Gas Export Monopoly Enters into Force', *Ria Novosti*, 2 December 2013. The new legislation leaves untouched Gazprom's monopoly on pipeline gas exports to Europe, while the other Russian companies can also work on the Kremlin's strategy to double by 2020 Russia's share in the global LNG market, which is only 4.5 per cent at present. Russia's only LNG plant at the moment is the Sakhalin project owned by Gazprom and the Royal Dutch Shell, which produces 10 million tonnes of gas annually.

13 'Putin Signs Order Challenging Gazprom Pipeline Monopoly, Report Says', *Moscow Times*, 22 July 2014. Even as Alexei Miller had said in May the previous year that the Gazprom had no plans to give other companies access to the pipelines, the Presidential Energy Commission effectively controlled by the Rosneft CEO Igor Sachin, the commission's executive secretary and a close ally of Putin, drew up the order for the government to let them participate in the pipeline use and construction. Rosneft had already threatened to take Gazprom to court if it was denied access to the planned 'Power of Siberia' pipeline that would link east Siberian gas production to China.

14 Vladimir Volkov, 'Gazprom: A Risky Strategy', *The Power of Oil and Gas* (Carnegie Moscow Centre), Vol. 10, nos 2–3. http://carnegie.ru/ proetcontra/?fa=24830. Accessed on 23 November 2013. The author adds that the initiatives by the Russian Ministry of Finance and the Ministry of Economic Development and Trade to skim off the Gazprom's super profits constituted an extra burden on the company.

15 Irina Mironova, 'A Sneak Peak into Russia's Energy Strategy upto 2035', *European Energy Review*, 30 January 2014. http://www.europeanener gyreview.eu/site/pagina.php?id=4250. Accessed on 13 January 2015.

RUSSIA

16 Akira Miyamoto, *Natural Gas in Central Asia: Industries, Markets and Export Options of Kazakhstan, Turkmenistan and Uzbekistan* (The Royal Institute of International Affairs, London, 1997), p. 77.
17 Sudha Mahalingam, 'India-Central Asia Energy Cooperation', in Santhanam K. and Ramakant Dwivedi, eds., *India and Central Asia: Advancing the Common Interest* (IDSA and Anamaya Publishers, New Delhi, 2004), p. 127.
18 Annon, 'Can Gazprom Be a Reliable Central Asian Gas Supplier to Europe?', *E-International Relations Students*, 5 July 2012. http://www.e-ir. info/2012/07/05/can-gazprom-be-a-reliable-central-asian-gas-supplier-to-europe/. Accessed on 3 August 2014.
19 'Global Energy: New Geopolitical Equations', *International Affairs* (Moscow), Vol. 58, no. 6, 2012, pp. 179–192.
20 Ibid.
21 Stratfor is based in Austin, Texas. In its mission statement, it calls itself a geopolitical intelligence firm that provides strategic analysis and forecasting to individuals and organisations around the world. It claims to provide global awareness and guidance to individuals, governments and businesses by using an intel-based approach to analyse world affairs. https://www.stratfor.com/about. Accessed on 6 April 2015.
22 Lauren Goodrich and Marc Lanthamann, 'A New Era for Russia's Energy Strategy?', *Stratfor*. Reprinted in *Economy Watch*, 14 February 2013. http://www.economywatch.com/economy-business-and-finance-news/a-new-era-for-russias-energy-strategy.14–02.html. Accessed on 13 March 2013.
23 Under the take-or-pay provision in a contract, the buyer agrees to pay a penalty in case he fails to buy as per the deal.
24 Vladimir Isachenkov, 'Russia Secures Ownership of Belarus Gas Pipelines', *AP*, 25 November 2011.
25 In comparison, the gas price for Gazprom customers in Europe hovers around $400 per thousand centimetres. 'Russia's Gazprom Takes over Gas Monopoly in Kyrgyzstan for $1', *Russia and India Report*, 11 December 2013.
26 Bruce Pannier, 'Gazprom Works to Advance Russia's Interests in Central Asia', *RFE/RL*, 20 October 2008.
27 During the Soviet times, the pattern was as follows: 'Good neighbours would come to Moscow, say nice things, Have their photos taken with our leaders, including the president, and we would lower oil and natural gas prices for them. All of that was very much to their liking, but very little to the liking of the Russian people.' V. Lukin, 'Energy: A Market Oriented Approach', *International Affairs* (Moscow), Vol. 54, no. 1, 2008, p. 73.
28 The text of the Strategy is available at http://www.energystrategy.ru/projects/docs/ES-2030_(Eng).pdf. Accessed on 9 January 2016. For a strident analysis of the Strategy, see Weitz, n. 4.
29 Michael Lelyveld, 'Gas Trader Itera Shows Signs of Trouble', *RFE/RL*, 19 October 2001.
30 Joshua Kucera, 'US Intelligence: Russia Sabotaged Pipeline Ahead of 2008 Georgian War', *EurasiaNet*, 10 December 2014. http://www.eurasianet.org/node/71291. Accessed on 3 March 2015.

111

31 Paul Goble, 'North-South Energy Routes More Attractive than East-West Ones', *Gas and Oil*, 24 January 2009. http://www.gasandoil.com/news/2009/02/ntr90740

32 Margarita M. Balmaceda, 'Ukraine's Energy Policy and US Strategic Interests in Eurasia', Woodrow Wilson International Centre for Scholars, Kennan Institute, *Occasional Paper 291*, May 2004. http://www.wilsoncenter.org/sites/default/files/op291_ukraines_energy_policy_balmaceda_2004.pdf. Accessed on 30 August 2013.

33 The Russian Black Sea fleet is based in Sevastopol. Though its population is predominantly Russian, Sevastopol was transferred to Ukraine in 1954 as a result of a re-drawing of internal boundaries in the USSR and it thus belongs today to the independent republic of Ukraine. In 1997, Russia and Ukraine concluded an agreement under which the former maintains its Black Sea fleet in Sevastopol on payment of a stipulated rent. Chandrashekhar Dasgupta, 'Ukraine, India and Energy Security', *Gas and Oil*, 18 January 2006. http://www.gasandoil.com/news/2006/02/nts60608

34 Jonathan Stern, *The Russian Gas Bubble: Consequences for European Gas Markets* (Royal Institute of International Affairs, London, 1995), pp. 56–62.

35 'Entering the New Millennium: Impact of Energy Price Development and Market Liberalisation 2002'. (*ECC Gas Centre Series No. 2*, United Nations, New York), p. 6.

36 www.gasandoil.com/goc/news/nte22840.htm. Accessed on 9 March 2005.

37 Stern, n. 34, p. 87.

38 Isabel Gorst, 'Broadening Export Strategy', *Petroleum Economist*, Vol. 71, no. 5, May 2004, p. 21.

39 Olga Oliker, 'Ukraine and the Caspian: An Opportunity for the United States'. (*Issue Paper*, Rand, Santa Monica, 2000). www.rand.org/publications/ip/ip198/. Accessed on 9 March 2005.

40 *Asia Times*, 1 March 2003.

41 Keith Smith, *Russian Energy Policies in the Baltics, Poland and Ukraine* (Center for Strategic and International Studies, Washington, DC, CSIS Report, December 2004).

42 N. J. Watson, 'A Blow for Caspian Exports', *Petroleum Economist*, Vol. 71, no. 10, October 2004, pp. 20–21.

43 The protesters against the Yanukovych victory, clad in orange, succeeded in getting a re-poll that Yushchenko won – hence the name 'Orange Revolution'.

44 Adrian Karatnycky, 'Ukraine's Orange Revolution', *Foreign Affairs* (New York), Vol. 84, no. 2, March–April 2005, p. 51.

45 Swaminathan S. Anklesaria Aiyar, 'Pipeline Lessons for India from Ukraine', *Times of India*, 7 January 2007.

46 13 April 2012. http://www.bloomberg.com/news/2012–04–13/ukrainian-parliament-backs-bill-to-ban-sale-of-gas-pipelines.html. Around the same time, Ukraine agreed to Gazprom's request to increase its daily transit capacity of Russian gas to the EU by 17.3 cm or about 7 per cent, reported the Ukrainian state gas pipelines and gas depots operator – Ukrtransgas. The request was associated with temporary repairs at the Russian

North Stream gas pipeline. 'Ukraine Helps Russia Fulfill Its Gas Supply Obligations to the EU', *Your Industry News*, 23 April 2012. http://www.yourindustrynews.com/ukraine+helps+russia+fulfill+its+gas+supply+obligations+to+the+eu_76744.html. Accessed on30 August 2014.

47 A powerful visual of the winter that year was that heavy snow fell in Rome for the first time in a quarter century and Venice's canals began to freeze. Robin M. Mills, 'Freezing in the Dark', *Foreign Policy*, 7 February 2012. http://www.npr.org/2012/02/07/146513127/foreign-policy-freezing-in-the-dark. Accessed on 7 April 2012.

48 Nicu Popescu, 'Ukraine's Gas Loop', *ISS Alert*, July 2013. http://www.iss.europa.eu/uploads/media/Alert_Ukraine_gas_01.pdf. Accessed on 3 March 2015.

49 Edward C. Chow, Sarah O. Ladislaw and Michelle Melton, 'Crisis in Ukraine: What Role Does Energy Play?', *CSIS*, 17 March 2014. http://csis.org/publication/crisis-ukraine-what-role-does-energy-play. Accessed on 3 March 2015.

50 'The Division of Europe That Marked the Cold War for 40 Years Would Now Reappear on Russia's Border: The New Division of Europe Would Be Right on Russia's Border: A NATO Ukraine and a Russia Ukraine, Two Nuclear Powers Facing Each Other on Russia's Border', Dan Rather Reports, *HDNet*, 14 July 2009.

51 *Business News Europe*, 1 Aril 2015.

52 The Ukrainians are confident of continuing the transit even after the expiry of the contract. Elena Kosolapova, 'Russia to Continue Its Gas Supply through Ukraine after 2018', *Trend News Agency*, 28 December 2015. http://en.trend.az/business/economy/2474808.html. Accessed on 28 December 2015.

53 'Ukrainian Parliament Backs Bill to Open Gas Pipelines to EU, US Firms', *Reuters*, 14 August 2014.

54 Peter Leonard, ' Ukraine Gets Loan to Modernise Gas Pipelines', *Associated Press*, 1 December 2014.

55 Parag Khanna, *The Second World: Empires and Influence in the New Global Order* (Allen Lane, London, 2008), p. 16.

56 Carlotta Gall and Thomas de Waal, *Chechnya: A Small Victorious War* (MacMillan, London, 1997), p. 127.

57 Hooman Peimani, *The Caspian Pipeline Dilemma: Political Games and Economic Losses* (Praeger, Westport, 2001), p. 55.

58 Ibid. Also see Michael T. Klare, *Resource Wars: The New Landscape of Global Conflict* (Henry Holt, New York, 2001), pp. 90–91.

59 See Elaine Holoboff, 'Oil and the Burning of Grozny', *Jane's Intelligence Review*, June 1995, pp. 253–257.

Also see Pavel Baev, 'Russia Refocuses Its Policies in the Southern Caucasus', *Working Paper, Caspian Studies Program*, 13 July 2001. http://belfercenter.ksg.harvard.edu/publication/3055/russia_refocuses_its_policies_in_the_southern_caucasus.html. Accessed on 4 December 2011.

60 Vicken Cheterian, *Dialectics of Ethnic Conflicts and Oil Projects in the Caucasus* (Programme for Strategic and International Security Studies, The Graduate School of International Studies, Geneva, 1997), p. 52.

61 *Energy Information Administration*, 12 March 2014. http://www.eia. gov/countries/cab.cfm?fips=RS

62 'Nord Stream', *Gazprom*, http://www.gazprom.com/about/production/ projects/pipelines/nord-stream /. Accessed on 8 April 2015.

63 Belarus has signed the 'Treaty on the Creation of a Union State of Russia and Belarus', with Russia in 2000, which envisages a federation between the two countries with a common constitution, flag, national anthem, citizenship, currency, president, parliament and army. In strict sense of the term, therefore, it cannot be considered a transit state.

64 'Nord Stream Can't Restrict Other Suppliers', *Pipeline and Gas Journal*, May 2012, Vol. 239, no. 5. http://www.pipelineandgasjournal.com/nord-stream-can%E2%80%99t-restrict-other-suppliers. Accessed on 3 March 2014.

The EU ruling follows the 'Energy Market Legislation' enacted to improve the functioning of the internal energy market and resolve structural problems. It covers five main areas: unbundling energy suppliers from network operators, strengthening the independence of regulators, establishment of the Agency for the Cooperation of Energy Regulators (ACER), cross-border cooperation between transmission system operators and the creation of European Networks for Transmission System Operators, and increased transparency in retail markets to benefit consumers. https:// ec.europa.eu/energy/en/topics/markets-and-consumers/market-legisla tion. Accessed on 3 March 2015.

In the present context, the relevant provision is the unbundling or separation of energy supply and generation from the operation of transmission networks. Premised on the principle that if a single company operates a transmission network and generates or sells energy at the same time, it may have an incentive to obstruct competitors' access to infrastructure, the provision seeks to attain a share of the Gazprom network of pipelines.

65 'Nord Stream II: Bypassing Ukraine and Dividing EU', *Fulton County News*, 13 November 2015. http://www.newsfultoncounty.com/econom ics/news/1315864-nord-stream-ii-bypassing-ukraine-and-dividing-eu. Accessed on 28 December 2015.

66 Barbara Lewis and Oleg Vukmanovic, 'Russia's Nord Stream Gas Pipeline Threatens EU Unity', *Reuters*, 16 December 2015. http://www.reuters. com/article/ukraine-crisis-nord-stream-idUSL8N1444MR20151216. Accessed on 28 December 2015.

67 Laurence Norman, 'Ukraine Prime Minister Calls on EU to Block Nord Stream II Pipeline Project', *Wall Street Journal*, 7 December 2015. http:// www.wsj.com/articles/ukraine-prime-minister-calls-on-eu-to-ban-nord-stream-ii-pipeline-project-1449493087. Accessed on 24 December 2015.

68 Quoted in Nick Cunningham, 'European Leaders Cry Foul against Germany's Support for Gas Pipeline', 21 December 2015. http://oilprice.com/ Energy/Natural-Gas/European-Leaders-Cry-Foul-Against-Germanys-Support-for-Gas-Pipeline.html. Accessed on 29 December 2015.

69 Chris Johnstone, 'Projected Russian Gas Pipeline Opens Rifts in Central Europe', *Radio Praha*, 9 December 2015. http://www.radio.cz/en/sec tion/marketplace/projected-russian-gas-pipeline-opens-rifts-in-central-europe. Accessed on 10 December 2015.

70 Alan Riley, *Wall Street Journal*, 16 December 2015. http://www.wsj.com/articles/challenging-putins-pipeline-strategy-1450297407. Accessed on 29 December 2015.

71 'Merkel Defends Nord Stream-2 Pipeline', *Russia Today*, 18 December 2015. https://www.rt.com/business/326440-merkel-gas-nord-stream2/. Accessed on 26 December 2015.

72 Cunningham, n. 68.

73 http://www.gazprom.com/about/production/projects/pipelines/south-stream. Accessed on 8 April 2015.

74 http://www.gazprom.com/press/news/2011/march/article110354. Accessed on 25 February 2015.

75 Österreichische Mineralölverwaltung. The Company changed its name to OMV in 1995 for the sake of simplicity.

76 According to the OMV's CEO Wolfgang Ruttenstorfer. Quoted in http://www.gazprom.com/press/news/2010/april/article97877. Accessed on 25 February 2015.

77 'South Stream Pushes into Troubled Waters', *Petroleum Economist*, June 2014, p. 49.

78 *Petroleum Intelligence Weekly*, Vol. 53, no. 28, 14 July 2014.

79 The EU alleges that Gazprom prevented countries from diversifying their gas supplies, broke anti-monopoly rules and imposed unfair prices calculated by a formula pegged to the price of oil that's opposed by some EU countries. *Global Post*, 15 September 2012. http://www.globalpost.com/dispatch/news/regions/europe/russia/120914/EU-Russia-probe-Putin-Gazprom. Accessed on 30 September 2012.
   The EU insisted that Russia sign and meet its obligations under the Energy Charter treaty. Russia did sign it but did not ratify it on the ground that it impinges on Russia's national interests by demanding that it give third party access to its pipelines. The EU–Russia talks have been going on. In March 2013, G. H. Oettinger, member of the European Commission for Energy, and A. V. Novek, minister of energy of the Russian Federation signed the road map for future cooperation in their capacities as the coordinators of the EU–Russia Energy Dialogue.

80 Bulgaria was set to reap $600 million per year in transit fees. 'Bulgaria Ready to Issue South Stream Permits', *Russia Today*, 19 December 2014. http://rt.com/business/215983-bulgaria-permit-south-stream/. Accessed on 30 December 2014.

81 'Vladimir Putin Said after Talks with His Turkish Counterpart, Recep Tayyip Erdogan', *Russia Insider*, 2 December 2014. http://russia-insider.com/en/politics_business/2014/12/02/09-55-23am/south_stream_no_more_hello_turk_stream. Accessed on 30 January 2015.

82 In December 2011, Turkey had already agreed to become a party to the original pipeline proposal, which Putin had called Ankara's 'wonderful Christmas gift to Moscow'. Yigal Schleifer, 'Turkey: In Pipeline Deal with Russia, Did Ankara Get a Bargain It Can't Afford?', *Eurasianet*, 11 January 2012. http://www.eurasianet.org/node/64823. Accessed on 12 February 2012.

83 Ibid.

84 Stephen Bierman, Ilya Arkhipov and Elena Mazneva, 'Putin Scraps South Stream Gas Pipeline after EU Pressure', *Bloomberg*, 2 December 2014.

https://www.bloomberg.com/news/2014–12–01/putin-halts-south-stream-gas-pipeline-after-pressure-from-eu.html. Accessed on 8 January 2015.

85 *Gazprom.* http://www.gazprom.com/press/news/2015/january/arti cle213570 /. Accessed on 8 April 2015.

86 The Blue Stream operated in spite of Russian suspicions that Turkey aided the Chechens during the 1994–1996 war, while Turkey has its own suspicions that Russia supported Turkey's secessionist the Kurdish Partiya Karkeren Kurdistan (PKK) movement.

87 The Baku–Tbilisi–Ceyhan (BTC) pipeline is a 2,000-km long crude oil pipeline from Baku in Azerbaijan to Ceyhan in Turkey. It was built to avoid the route passing through either Russia or Iran. The US energy secretary Bill Richardson witnessed intergovernmental agreement signed at a meeting of the Organisation for Security and Cooperation in Europe in Istanbul on 18 November 1999, where he said, 'This is not just another oil and gas deal, and this is not just another pipeline. It is a strategic framework that advances America's national security'. Quoted in Klare, n. 58.

88 Turkey's inflated demand estimates resulted in it having to pay penalties to Iran and Russia for the natural gas it was unable to obtain, which came to $1 billion in 2002. The Iranians did not believe the demands were inflated. They said that the demands were, instead, met from a discounted price from Russia. Sensing problem, Ankara rushed to reach an agreement on building a pipeline to Greece so that it could pass on its surplus. In March it signed a protocol with Greece to build a 285-kilometre line linking the two Mediterranean rivals to ease the pressure of oversupply. Alec Rasizade, 'The Caspian Energy Legend and the "Great Game" of Concomitant Pipelines', *Iranian Journal of International Affairs*, Vol. 14, nos 1–2, Spring–Summer 2002, p. 40.

89 Gulshan Dietl, 'Will Turkey Be the New Hub for Gas', *IDSA Comments*, 6 November 2013. http://www.idsa.in/idsacomments/WillTurkeybethene whubforgas_gdietl_061113.html

90 Kelly Gilblom, Elena Mazneva and Anna Shiryaevskaya, 'Ten Billion Reasons Why Russia Will Balk at Curbing Turkey's Gas', *Bloomberg News*, 25 November 2015. http://www.bloomberg.com/news/arti cles/2015–11–25/ten-billion-reasons-why-russia-will-balk-at-curbing-turkey-s-gas. Accessed on 30 November 2015.

91 Quoted in Denis Pinchuk and Olesya Astakhova, 'Russia May Freeze Turkish Stream Gas Project: Gazprom Sources', *Reuters*, 1 December 2015. http://www.reuters.com/article/us-mideast-crisis-russia-tur key-gas-iduskbn0tk4wy20151201#02mkhhoevshofx2i.97. Accessed on 5 December 2015.

92 'Russia Says Talks Suspended with Turkey on Turkish Stream Pipeline', 3 December 2015. http://www.hurriyetdailynews.com/russia-says-talks-suspended-with-turkey-on-turkish-stream-pipeline-.aspx?pageID=238&n ID=92022&NewsCatID=348. Accessed on 5 December 2015.

93 'Turkey Has Shelved Turkish Stream Gas Pipeline Project, Says President Erdoğan', 5 December 2015. http://www.hurriyetdailynews.com/turkey-has-shelved-turkish-stream-gas-pipeline-project-says-president-erdogan.aspx?pageID=238&nID=92115&NewsCatID=348. Accessed on 6 December 2015.

94 'Putin: Turkish Stream Pipeline Project to Happen If Turkey Gets Guarantees from Brussels', *The TASS*, 17 December 2015. http://tass.ru/en/politics/845011. Accessed on 30 December 2015.
95 'Gazprom Delegation Visits China', 15 September 2009. http://www.gazprom.com/press/news/2009/september/article68067/
96 'Gazprom to Sign Monumental Gas Deal with China', 19 May 2014. http://rt.com/business/159880-gazprom-china-russia-cnpc/
97 Chang Felix K., 'Friends in Need: Geopolitics of China-Russia Energy Relations', *Foreign Policy Research Institute*, May 2014. http://www.fpri.org/articles/2014/05/friends-need-geopolitics-china-russia-energy-relations
98 'Russia–China Sign $400 Billion Natural Gas Deal', *Economy Watch*, 21 May 2014. http://www.economywatch.com/news/russia-china-400-billion-natural-gas-deal.22–05.html. Accessed on 21 May 2014.
In the end, the two sides reportedly agreed to a price that was closer to China's bid than Russia's ask. Elena Mazneva, Aibing Guo, and Benjamin Haas, 'Russia Close to $400 Billion Gas Pipeline Deal in Pivot to China', *Bloomberg*, 19 May 2014; Abheek Bhattacharya, 'Ukraine Gives China an Edge on Russian Gas', *Wall Street Journal*, 15 May 2014.
Russia Today also expected the price to be agreed at between $350–400 per thousand cubic metres. 'Gazprom to Sign Monumental Gas Deal with China', 19 May 2014. http://rt.com/business/159880-gazprom-china-russia-cnpc/. Accessed on 21 May 2014. Gazprom's average price in Europe was $380.5 per thousand cubic metres last year. The price in today's contract is more than $350, Interfax reported, citing a person it didn't identify.
99 Elena Mazneva and Stepan Kravchenko, 'Russia, China Sign $400 b Gas Deal after Decade of Talks', *Bloomberg*, 21 May 2014. http://www.bloomberg.com/news/articles/2014–05–21/russia-signs-china-gas-deal-after-decade-of-talks
100 The region has become critical for the future viability of Russian energy sector as production declines in West Siberia, whose giant fields were discovered first and are much closer to Europe. Jane Nakano and Edward C. Chow, 'Russia-China Natural Gas Agreement Crosses the Finish Line', *CSIS*, 28 May 2014. http://csis.org/publication/russia-china-natural-gas-agreement-crosses-finish-line. Accessed on 30 June 2014.
101 'Vladimir Putin's Meeting with Heads of Leading International News Agencies', 24 May 2014. http://eng.kremlin.ru/news/7237. Accessed on 30 June 2014.
102 Thomas R. Stauffer, 'Caspian Fantasy: The Economics of Political Pipelines', *Brown Journal of World Affairs*, Vol. VII, no. 2, Summer–Fall 2000, pp. 63–78.
103 Charles Krauthammer, 'Who Made the Pivot to Asia? Putin', *Washington Post*, Reprinted in *The Hindu*, 24 May 2014.
104 Gordon Chang, 'China and Russia Form an Enduring Partnership', *China US Focus*, 28 May 2014. http://www.chinausfocus.com/foreign-policy/china-and-russia-form-an-enduring-partnership/. Accessed on 3 January 2015.
105 Steven W. Lewis, 'Russia to China Gas Deal: A Game of Spigots?', *Forbes*, 22 May 2014. http://www.forbes.com/sites/thebakersinstitute/2014/05/22/

russia-to-china-gas-deal-a-game-of-spigots/. Accessed on 30 May 2014. There were strong rumours that Beijing wanted to relieve the virtual duopoly CNPC and Sinopec of their private pipeline networks and transfer control to an overarching national pipeline firm with private participation. 'Beijing Mulls New Pipeline Ownership Plan', *Petroleum Economist*, Vol. 53, no. 12, 24 March 2014, p. 2.

106 Ibid.

107 Meidan Mical, 'The Implications of China's Energy-Import Boom', *Survival*, Vol. 56, no. 3, June–July 2014, p. 187.

108 Franz-Stefan Gady, 'China to Receive Russia's S-400 Missile Defense System in 12–18 Months', *The Diplomat*, 17 November 2015. http:// thediplomat.com/2015/11/china-to-receive-russias-s-400-missile-defense-systems-in-12–18-months/. Accessed on 5 January 2016.

109 Henry Meyer and Evgenia Pismennaya, 'Putin Deals China Winning Hand as Sanctions Power Rival', *Bloomberg*, 13 October 2014.

110 Lucy Hornby, 'Putin Snubs Europe with Siberian Gas Deal that Bolsters China Ties', *Financial Times*, 10 November 2014.

111 Quoted in Alexei Anishchuk, 'Russia China Ink Framework Deal', *Reuters*, 9 November 2014.

112 'Russian Draft Energy Strategy Sees 23% of Exports to Asia-Pacific By 2035', *Platts*, 24 January 2014. http://www.platts.com/latest-news/natural-gas/moscow/russian-draft-energy-strategy-sees-23-of-exports-26649363. Accessed on 23 May 2014.

113 Nakano and Chow, 'Russia-China Natural Gas Agreement Crosses the Finish Line, n. 100.

114 The US, Canada, the EU and Japan have selected their own preferred ways of imposing the sanctions: the US has chosen to freeze the Russian assets; ban on the energy firms like Novatek, Rosneft and Gazprombank; and travel bans. The EU has been more circumspect, seeking to minimise adverse consequences for the civilian population or for legitimate activities.

115 According to a simplistic analysis, the fundamental driver of the deal was anti-Americanism meant to show that they are still a great power that can play against the US. Stephen Blank, 'Russia-China Natural Gas Deal Likely to Reshape Energy Markets', *Transcript of a Discussion on NPR*, 29 May 2014. http://northernpublicradio.org/post/russia-china-natural-gas-deal-likely-reshape-energy-markets. Accessed on 31 November 2014.

116 Established in May 2014, its member-states to date are Russia, Kazakhstan, Belarus and Armenia.

117 Dmitri Trenin, 'From Greater Europe to Greater Asia: The Sino-Russian Entente', *Carnegie Paper, Carnegie Moscow Center*, 9 April 2015.

118 Gulshan Dietl, 'Middle Kingdom and the Middle East: The Energy Connection', *China Brief*, Vol. 3, no. 4, 25 February 2003; *Jamestown Foundation*. http://www.jamestown.org/single/?tx_ttnews%5Bswords%5D=8fd5893941d69d0be3f378576261ae3e&tx_ttnews%5Bany_of_the_words%5D=Gulshan%20Dietl&tx_ttnews%5Btt_news%5D=20612&tx_ttnews%5BbackPid%5D=7&cHash=93afdb9c58c7936abb84ecbc6049ec89#.VSyw6_mUeHA

119 Ole Odgaard and Jorgen Delman, 'China's Energy Security and Its Challenges towards 2035', *Energy Policy* (Amsterdam), Vol. 71, 2014, pp. 107–117.

120 Felix K. Chang, 'Friends in Need: Geopolitics of China-Russia Energy Relations', *Foreign Policy Research Institute*, May 2014. http://www.fpri.org/articles/2014/05/friends-need-geopolitics-china-russia-energy-relations

121 Sergey Pravosudov, 'Lack of Russian Gas Strangles China', *Gazprom Magazine*, no. 6, http://www.gazprom.com/press/reports/2013/china-suffocates

122 M. K. Bhadrakumar, 'Russia Sheds Its Fear of China', *Indian Punchline*, 28 October 2013. http://blogs.rediff.com/mkbhadrakumar/2013/10/28/russia-sheds-its-fear-of-china/. Accessed on 30 October 2013.

123 MacDonald and Wimbush, 'India's Energy Security', n 2.

124 M. K. Bhadrakumar, 'Russia-China Take a Leap of Faith', *Indian Punchline*, 15 June 2015. http://blogs.rediff.com/mkbhadrakumar/2015/06/15/russia-china-take-a-leap-of-faith. Accessed on 23 June 2015.

5

TURKMENISTAN

Pawn and player in the game of chess

Turkmenistan is at present a high-value chip in the New Great Game, an energy-rich nation with a lot to offer and no clearly defined geopolitical orientation. As it continues to open up to the outside world, the fight for Turkmenistan will be an extremely important, if not decisive, factor for victory in the geopolitical conquest of Eurasia. While Turkmenistan is not in itself even a regional power, it is still in an advantageous position vis-à-vis external actors seeking to control Central Asia's energy deposits. Recent actions have shown that it is a willing energy player and will work with whoever grants it the best deal.[1]

Military escalation under the extended Af-Pak war bears a relationship to TAPI. Turkmenistan possesses third largest natural gas reserves after Russia and Iran. Strategic control over the transport routes out of Turkmenistan have been part of Washington's agenda since the collapse of the Soviet Union in 1991.[2]

Unlike Russia and Iran, Turkmenistan is not a decisive actor on the world scene. A reclusive state with a population of five million and a short history of a few decades as an independent entity, it has been more a pawn than a player at the chess-board – to resort to a much used, misused and tired metaphor of the 'Great Game'. The previous quotes refer to it as a contested territory, a reason for extending war in Afghanistan and a factor deciding the future of Eurasia. The chapter proposes to bring out Turkmenistan's own policies and preferences as reflected in its decisions on the rich endowments of its energy resources with special reference to the gas pipelines.

Turkmenistan's gas

The Turkmen geography is less than fortunate. It is a landlocked country; bounded by the Caspian Sea to its west and the states of Kazakhstan, Uzbekistan, Iran and Afghanistan. The Caspian Sea is landlocked itself. Its division among the littoral states and its legal status are yet to be finalised. It is also a great impediment to the energy passage. Certain types of floating production platforms for deep-water exploration and exploitation are far too big to move into the Caspian. Such equipment hardware is generally not available locally, since production platforms usually are built with parts from different countries. This means high costs for rigs and vessels. There are thus significant logistical constraints, and cycle times in exploration and exploitation are long.[3] In addition, there is recurring ice in the north of the Caspian, extreme depth differences from 5 to 1,000 metres and a high level of earthquake activity throughout the region.

The Turkmen gas is considered high in hydrogen sulphide and carbon dioxide having greater pressure and temperature. The quality is thought to be continuously deteriorating becoming sourer in terms of sulphur content. These factors pose technical challenges requiring greater capital costs for exploration and development. It also increases difficulty in maintaining the wells, and raises the costs of removing impurities to provide pipeline quality gas.[4] The reserve size, gas quality, pipeline capacity and availability, commitment of the leadership to honouring contracts and relations with the neighbouring countries are listed as the factors that make the gas production and sales fraught with uncertainties.[5]

Complicating Turkmenistan's ability to move its gas in the future is the stated goal of both Kazakhstan and Uzbekistan to increase their gas exports using the Central Asia Centre (CAC)[6] pipelines system. How much Turkmenistan can or will ship will be determined not only by pipeline capacity but by its neighbours' rights as well. Uzbekistan has announced, for example, that it plans to export some of its gas on the CAC system, and would therefore make only 20 bcm per year of export space available to Turkmenistan.[7]

Its two main oil and gas-producing areas are the South Caspian basin and the Amu Darya basin. The existing oil fields are concentrated in the former region, whose geological structure lies along the Apsheron Still from the Apsheron Peninsula in Azerbaijan to the Celeken Peninsula in Turkmenistan. On the Turkmen side of the structure, oil production started at the beginning of the twentieth century and relatively large fields such as Celeken, Kotur-Tepe, Nebit-dag and Barsa Gelmes were discovered. Most natural gas production is concentrated in the Amu Darya basin, where the giant fields of Dauletabad 1.3 trillion cubic metres (tcm) and Shatlyk 1.0 tcm were discovered.

During the Soviet era, Turkmenistan was the second-largest natural gas supplier in the Soviet Union after Russia. For example, in 1990 Russia supplied 92 billion cubic metres (bcm) and Turkmenistan 78.7 bcm to the other republics and exported 96.0 bcm and 13.0 bcm, respectively, to Europe.[8] The Dauletabad gas field was previously believed to be the largest gas field in Turkmenistan with some 1.4 tcm. Most of Turkmenistan's agreements with partners were based on Dauletabad being the major contributor to gas pipelines. Till mid-2000, the proven reserves of natural gas were estimated to be large (around 3 tcm) and unexplored reserves to be huge (12–21 tcm).[9] Extrapolating from the Soviet geological data, some analysts estimated the reserves at 10 tcm to 14 tcm, ranking the country fourth in gas assets.[10]

The continuous upgrading of data saw a sudden spurt in 2008, when the Turkmen government hired the British firm Gaffney, Cline, and Associates (GCA) to survey some of the country's gas fields. The survey delivered a report that went beyond any guesses made around that time. On 13 October, GCA manager Jim Gillet announced that the results of an audit from one field, the South Yolotan–Osman field in south-eastern Turkmenistan, showed it to be a 'world class field' with a low estimate of 4 tcm, a best estimate of 6 tcm and a high estimate of 14 tcm.[11]

Gillet elaborated on the Yolotan find later at a conference in London. He called it a 'Super Giant' field and one of the largest undeveloped deposits. The field, he said, is 75 kilometres (km) long and 35 km wide, has a gas column of 500 metres and is likely to hold 6 tcm of gas.[12] Eleven wells have been drilled in South Yolotan–Osman; seven of which have fully penetrated the reservoir. At least seven parallel plants are possible, each producing 30 cm gas a day; one of them will produce well over 1,000 tons of sulphur a day. Turkmenistan plans to develop the field in a series of phases each able to produce 10 bcm a year, he said.

Turkmenistan's Yashlar deposit is also a 'very significant' field, Gillett said. Close to the South Yolotan–Osman deposit, Yashlar holds about 700 bcm of gas at a 'best estimate'. It may contain as much as 1.5 tcm, he added. Four wells have been drilled at Yashlar so far, only one penetrated the reservoir.[13]

A few years later, the GCA revised the audit further, enhancing the estimates of the Yolotan field. Peter Holding, the GCA general manager for Russia and the Caspian, told an energy conference in 2011 that the field was the second-largest field after the South Pars field in Iran and that new seismic data and wells drilled had proven that South Yolotan and the adjacent Osman field were actually a single structure. The Turkmen government officials attending the conference said that they had estimated the reserves at 21 tcm.[14]

Gazprom deputy chairman Alexander Medvedev immediately dismissed the findings of the audit firm. He countered the estimates by saying that the Soviet Union had already carried out geological surveys according to which the gas deposits were much smaller. Since then, two more publications have cast doubt on the audit results, relying on information obtained from unnamed Russian and Turkmen sources who suggested that Turkmen officials may have provided false data to exaggerate the size of the reserves. Gaffney, Cline & Associates refuted these allegations. Meanwhile, President Gurbanguly Berdimuhamedov dismissed various top energy officials,[15] creating further mystery around the entire episode.

In November 2011, Berdimuhamedov issued a decree, according to which the super-giant oil and gas area, including Southern Yolotan–Osman, Minara and other fields, was renamed as 'Galkinish', which in the Turkmen language means Revival.[16]

Gas policies

In 1995, Turkmenistan proclaimed a policy of international diplomatic neutrality, which was unanimously recognised by the United Nations General Assembly in December 1996. The energy policies were framed and executed within this broad context. It was a move to assert its independence vis-à-vis Russia. When Putin sought to persuade the three gas-rich Central Asian states – Kazakhstan, Turkmenistan and Uzbekistan – to unite with Russia to form a 'Gas Alliance',[17] the then president Saparmurat Niyazov was the most outspoken in rejecting the suggestion. He dismissed it as 'nonsense till the ways to the gas markets and the membership of alliance and price policy is fixed'.[18] What he meant was that Russia should bring down the transit fees for the Central Asian gas via Russia to Europe. He maintained that his country was satisfied with the system of selling gas at $42 per thousand cm on the Turkmen–Uzbek border. 'We are already experienced, and are willing to sell our gas only at the border.'[19] Selling gas at the border has since remained the oft-quoted mantra of the country's gas policies.

On the occasion of the fifth anniversary of the country's neutrality, Niyazov gave a televised speech titled 'Turkmenistan: Neutrality, History, Outlook and National Strategy', in which he said that transferring gas across the Turkmen soil to the international markets was the most economic alternative for Turkmenistan. He went on to dilate on 'certain countries' political intrigues, disguised as economic projects that were actually meant to create obstacles in the way of transferring Turkmenistan's oil and gas to the international markets'. Referring to the level of Turkmenistan's gas exports before independence, he regretted that although the country brought in some $15 to $20 billion annually, the proceeds were not spent for its prosperity.

By that time, he had already declined the Russian invitation to join the gas alliance. In the televised address, he proceeded to explain the reasons for Turkmenistan downgrading its membership to an associate member in the Commonwealth of Independent States (CIS)[20] – also a Russian initiative. He said that 'with bitter experience of previous futile political alignments', the country should avoid joining any new alliances, particularly military ones. Turkmenistan had sufficient military power and it had no border disputes with its neighbours, he said. He was satisfied with the regional cooperation within the framework of the Economic Cooperation Organisation (ECO)[21] of which the country was a member.[22]

At the beginning of July 1996, the oil and gas sector was reorganised by a presidential decree in order to activate the stagnating industry. The former Ministry of Oil and Gas was abolished and the Ministry of Oil and Gas Industry and Mineral Resources took its place. The new ministry incorporated Turkmen Geology, which had been directly subordinate to the cabinet ministers as an independent organisation. It had three main objectives: first, in-depth economic analysis of the oil and gas industry; second, long-term strategic planning and investment (including foreign investment); and third, development and transfer of scientific and new technology.[23] The newly formed 'Turkmengeologiya' was tasked with exploration and prospecting of hydrocarbon deposits in the country. It was mandated to carry out two-dimensional and three-dimensional seismic survey, gravitational exploration, geologic–geochemical methods of investigation, as well as deep prospect and parametric drilling aimed at finding of new accumulations and growth of hydrocarbon reserves.[24]

Niyazov passed away in December 2006. Gurbanguly Berdimuhamedov succeeded him.[25] The energy sector under his presidency has been a priority sector. Steps have been taken on many fronts to develop and modernise it. Soon after taking over, he ordered the vice prime minister Tachberdy Tagiev to conduct an audit of the country's hydrocarbon deposits, which 'will allow us to clarify our strategy for further developing the use of our hydrocarbon resources and account for the high goals set in the program of development of Turkmenistan's oil and gas industry through 2030'.[26] Simultaneously, the country began implementing a programme of modernising the infrastructure of the interstate pipelines, setting into operation new compressor systems equipped with modern control and management system. One billion dollars were spent for the purpose.[27] A target was set to more than treble the gas output to 230 bcm annually by 2030, of which 180 bcm would be exported.[28] Turkmen state agency for oil and gas resources opened an office in London with the hope to 'become an exporter to the whole world' and seeking foreign involvement in the development of its offshore energy reserves.[29]

A major statement on gas policy was made in August 2012, when Berdimuhamedov called a meeting of his energy officials, where he announced a detailed plan of export diversification. The most important in his statement was a conspicuous absence of any reference to Russia. The European Union (EU)–sponsored Nabucco project was also missing in the pronouncement. China, on the other hand, came in for a positive mention as his preferred choice of accepting loan. He made it clear that in terms of the export routes, he was looking south. The officials were instructed to 'take all necessary measures' to ensure that exports through Turkmenistan–Afghanistan–Pakistan–India (TAPI) would be arranged by the end of the year. He made a surprise announcement that the Italian company Eni[30] would become the second foreign company – after the China National Petroleum Corporation (CNPC) – to be given the right to develop the Nebit-Dag oil and gas onshore field in western Turkmenistan.[31]

Turkmenistan is concerned about its unstable neighbourhood, particularly the security of its energy transport through it. It shares a long land border with the conflict-ridden Afghanistan and the Caspian maritime frontier, where the issues of legal regime and territorial divisions are not finalised. The country has been in a long-winding dispute with Azerbaijan at the other end of the Caspian. Its energy policies reflect these concerns. At the United Nations General Assembly, it co-sponsored a resolution on 'Reliable Energy Transit', which was passed on 19 December 2008. The resolution recognised the need for international cooperation to ensure 'the reliable transportation of energy to international markets through pipelines and other transportation systems'. Next year in April, the country hosted a conference on the topic.[32] The explosion on its pipeline with Russia underscored the necessity of a multi-lateral solution to the problem. On 30 September 2013, Deputy Prime Minister Rashid Meredov spoke at the United Nations General Assembly (UNGA), expressing a desire to increase 'constructive and omnilateral [sic] cooperation with the UN' and suggesting the creation of a UN 'Universal International Law Tool Kit' to promote global energy security. He offered to host several international conferences of energy experts towards that end.[33]

Turkmen search for security went beyond the UN all the way to North Atlantic Treaty Organisation (NATO). As a self-proclaimed neutral country, Turkmenistan broke with its history to become the first Central Asian state to join the NATO 'Partnership for Peace' programme.[34] Berdimuhamedov went to Bucharest for the NATO summit in 2008. The Summit paved the way for the Euro-Afghan delivery corridor for non-lethal cargo. At his meetings with the US President George W Bush, he was reported to have discussed various gas pipeline proposals for his country. He also met the Afghan President Hamid Karzai to review the TAPI pipeline.

Gas pipelines

Turkmenistan has an unfortunate geography, as already noted. It is land-locked with a shoreline on the Caspian Sea, which is also landlocked. Its neighbourhood is extremely inhospitable to any kind of cross-border infra-structure. Since the country is far away from open seas, its only passage is via the pipelines. The vulnerable pipelines, the neighbours that are them-selves gas exporters, the remoteness of viable markets are serious limitations on the Turkmen pipeline proposals.

A prescriptive analysis suggests four principal options as far as exports from the region are concerned: first, selling gas to Russia at the border; second, selling gas to other Commonwealth of Independent States (CIS) countries at their borders as long as Russia permits Central Asian gas to pass through its territory; third, persuading Russia to carry gas to Europe for them; and, finally, shutting in their gas until such time as new export routes have been established. Since independence, first and second options have proved to be the only viable ones, but have not been satisfactory for the Central Asian countries. Their best option is the third one, namely, selling outside the CIS through Russia, but Gazprom has not permitted transit. The fourth option – waiting for new export routes to be established – could bring them substantial revenues, but only in the longer term.[35] It concludes that a compromise – recognising its dependence on Russia, no matter how difficult – is the most practical option. Using the Russian natural gas net-work which forms the hub of the whole CIS system will be unavoidable for the time being.[36]

A second opinion on the matter concurs: as long as major new pipelines are not in place, Turkmenistan's gas production and exports will continue to grow only sporadically, mirroring the payment abilities of its neighbours and its transit relationship with Russia.[37]

Things have evolved since these projections, as the Chinese entry has been a game changer in the availability of options for the gas exports from Central Asia. The Russian predominance continues, nonetheless. The Turkmenistan–Russia–Ukraine pipeline illustrates this point beyond doubt. The pipeline is the most contested not just among the three participants, but pulls in all the major global powers. The contest involves much more than just the gas transit. In fact, the pipeline unfolds the changing contours of international relations. The pipeline, accordingly, deserves an in-depth scrutiny.

Turkmenistan-Russia-Ukraine Pipeline

The Ukraine–Russia–Turkmen gas pipeline, till the break-up of the Soviet Union, was a domestic grid. In the post-Soviet era, all of Turkmenistan's gas exports outside Central Asia passed through Russia, which put the latter in near complete control of around three quarters of Turkmenistan's exports. Thus in the mid-1990s Gazprom, as the sole owner of the pipeline that took Turkmen gas to Europe via Russia, halted Turkmen exports and demanded an increase in transit fees. Turkmenistan had little choice but to agree, as efforts to build viable export routes via Afghanistan and Iran had largely failed.[38]

In April 2003, Niyazov signed a 25-year gas deal with Russia to export up to 80 bcm a year until 2028. Turkmenistan would get $44 per thousand cubic metres (cm) – half of it in cash and the rest in commodities. The market analysts saw it as a terrible deal as Russia got $100 per thousand cm for its gas in the European markets. 'The deal clearly short-changed Turkmenistan's future so that Niyazov could further consolidate his position at home.'[39] Alongside, negotiations were going on over the financial terms of a reconstruction project for the 3,940 km of the line within the Republic's borders. Turkmenistan, which had ambitious plans for increasing gas exports to Russia, was encouraging construction of a new export line that would skirt along the east coast of the Caspian Sea on its way north.[40] In July, Russia and Ukraine established a new company to take over the purchasing and transporting of Turkmen gas to Ukraine. RosUkrEnergy, as the company was named, was a further step by Russia to exert its control over the gas network involving the three states. In September 2006, Gazprom won access to the large Yolotan field and an option on any surpluses until 2009.[41]

In early May 2007, two rival energy summits promoted two different strategies for the export of Caspian energy resources. A meeting in Krakow on May 11 brought together the leaders of Azerbaijan, Georgia, Lithuania, Poland and Ukraine to sign a joint declaration calling for increased cooperation in transporting energy resources from the Caspian region to Europe via the Caucasus. The following day, Russian President Vladimir Putin eclipsed the Krakow meeting by securing the participation of Kazakh President Nursultan Nazarbayev, who had previously confirmed his presence in Krakow, to a parallel summit with Berdimuhamedov. The three leaders signed a major energy deal on the expansion of the Prikaspiisky or Pre-Caspian pipeline[42] to transport Turkmen natural gas to Russia along the Caspian Sea coast via Kazakhstan.[43]

As it was about to be implemented, a major pipeline that carried Turkmenistan's vast natural gas resources to Russia exploded in a massive fireball on 9 May 2009. A conspiracy theory put down the explosion to the April 2009 visit of Berdimuhamedov to Moscow, where he shocked his hosts by declining to sign agreements finalising arrangements for bilateral cooperation on refurbishing and reconstructing the Turkmenistan segment of the 'Prikaspiiskii'. The trilateral agreement 2007 had reserved to each of the three participating countries full autonomy and responsibility for carrying out the works on its own national territory.[44] It needs to be repeated here that according to Turkmen export policy, it sells its gas at the border. Therefore, under the contract, Turkmenistan supplies natural gas within the territory of the country. Russia purchases gas from the Daryalyk measurement point. It is the Gazprom that is responsible for further transportation and transit.

As was to be expected, a bitter blame game ensued. On the same day, Anatoly Dmitrievsky, director of the Institute for Oil and Gas Issues at Russia's Academy of Sciences, appeared on the Russian television programme Vesti. In comments that were widely publicised by Russian print media, Dmitrievsky blamed Turkmenistan's aging pipeline system for the blast, saying it was 'built in the late 1960s and start of the 1970s, is rather worn and in need of repair and reconstruction'. Dmitrievsky said it could also have been the fault of Turkmen dispatchers monitoring the pipeline. The Turkmen government website (Turkmenistan.ru) responded to Dmitrievsky's comments the following day, saying the Russian academic's comments 'did not correspondent to reality', and rejected it as an 'attempt to negatively portray the work of the dispatcher service of Turkmenistan'. It said that the blast was caused by a reckless and irresponsible decrease in the amount of gas drawn from the pipeline by Gazprom.[45]

Berdimuhamedov went further. He ordered the deputy prime minister for Oil and Gas Tachberdi Tagiev to meet with the Gazprom officials and outlined Turkmenistan's next steps if the talks fail. 'Unless Gazprom admits its faults, then we will invite independent experts to examine the causes of the blast,' he said. 'If we are responsible for the disorder, then we will fix the damages. If Gazprom is responsible for the blast, then they have to pay for the repairs and compensate Turkmenistan for pipeline damages.'[46] Although Gazprom charged that the explosion was the result of the pipeline's age and Turkmens' negligent maintenance, and the Turkmen government accused Gazprom of having sharply and unilaterally reduced gas volume, most observers understood the explosion in the context of the ongoing battle over gas pricing and transit volume between the two countries. Both versions likely contain elements of truth. The pipeline system was old and in relatively poor repair, but it was also the case that with less foreign demand, Gazprom was drawing off lower volumes of Turkmen gas than planned, and frequently did this with little forewarning.[47]

It seemed to have pushed Turkmen leaders to finally decide in favour of constructing the Trans-Caspian Pipeline (TCP) to Azerbaijan, with resources flowing from there to Turkey and potentially Western markets. Berdimuhamedov reportedly called Azerbaijan's president Ilham Aliyev to convey that message. The hint could also not have been clearer on the government's Website: Turkmenistan now intends 'to diversify supply of Turkmen natural gas and build the reliable and stable system to transit for Turkmen energy to international markets'.[48] The TCP line would keep Russia and Iran both out of the loop in its advance to the European markets. Both insist that the Caspian pipeline would be an environmental disaster, as Caspian is an inland water mass with no outlet. They invoke the legal stipulation that no decision can be taken unless there was a consensus among the Caspian littoral states – Russia, Iran, Kazakhstan, Turkmenistan and Azerbaijan.

China also does not want the Turkmen gas reaching Europe, as Turkmenistan would eventually demand the European price from China as well. Russia does not want to lose its market share in Europe to Turkmenistan. Moscow and Beijing are on the same side of the fence on the TCP although Russia is more belligerent.

There are threatening sounds against Turkmenistan joining the TCP. Konstantin Simonov, Director of the Fund for National Energy Security in Moscow, conjectured with obvious relish that 'only the experience of August war in Georgia is deterring Ashgabat today', adding that the fact that TCP has not begun yet after 15 years of talks indicates that the West has not given Ashgabat a 100 per cent guarantee of defence against the Russian military reaction.[49] In the circumstances, Turkmenistan is hedging its bets between Russia, China, TCP and Turkmenistan–Afghanistan–Pakistan– India (TAPI). In the meanwhile, a trilateral initiative involving Turkmenistan, Azerbaijan and Turkey has again been proposed in March 2015 after a long hiatus.

The fate of the Turkmen-Russian pipeline made one more flip-flop in a quick succession after that. In December 2009, an agreement in principle was announced in Ashgabat between presidents Berdimuhamedov and Dmitry Medvedev of Russia, according to which the two countries would cooperate in refurbishing and renovating a separate but related pipeline, the 'East–West Pipeline' that crosses the southern region of the country to terminate not far from its Caspian Sea coast. In May 2010, following a review of more than 70 international replies to a tender offer for that work, Berdimuhamedov announced that Turkmenistan had decided to execute itself the refurbishment and renovation of the East-West Pipeline. With that, the Russian preference of linking it to its own Central Asia Centre Pipeline came to naught. The line, built entirely by Turkmenistan, has been completed in late December 2015 that traverses from Galkinish all the way to the Caspian shore. Its onward direction is still not decided.[50]

Gas supplies from Turkmenistan to Russia resumed in January 2010. Since then, there are signs that the country is developing multiple export routes: north to Russia, east to China, south to Pakistan and India via Afghanistan and west to Europe via the Caspian Sea.[51]

Parallel gas lines to China

Niyazov's aversion to Russia led him to seek an export outlet through China. He arrived in China on a rare visit on 2 April 2006 and signed a framework agreement the next day. The text of the agreement was published by the Turkmen News Agency TDH according to which China would buy 30 bcm of gas each year for 30 years, starting in 2009. The Turkmen television provided additional information about the route. The station reported:

> In the first phase, we plan, starting from 2008, to deliver some 30 bcm of Turkmen gas via Uzbekistan and Kazakhstan, to Urumci [western China] and beyond it, to Shanghai [eastern China], and to increase these volumes to up to 50 bcm by 2010.[52]

The original agreement called for a price of $80 per 1,000 cm.[53]

The first move led to others in quick succession. On 30 April 2007, Uzbekistan and China signed an agreement on the construction and exploitation of the pipeline's Uzbekistan section. In July 2007, CNPC signed a production-sharing contract to explore and develop gas fields on the right bank of the Amu Darya River with the Turkmen State Agency, and a natural gas purchase and sales agreement with Turkmengazi State Concern.[54] On 8 November 2007, Kazakhstan's oil company KazMunayGas signed an agreement with the CNPC on principles of future work on the pipeline. The pipeline was inaugurated on 14 December 2009 in a ceremony in Saman-Depe during the Chinese president Hu Jintao's visit to Turkmenistan with the leaders of Turkmenistan, Uzbekistan and Kazakhstan.

The Central Asia-China Gas Pipeline starts at Turkmen-Uzbek border city Gedaim and runs through central Uzbekistan and southern Kazakhstan before reaching Horgos in China's Xinjiang Uygur Autonomous region. It is the longest line stretching 7,000 km, which is over one-sixth the circumference of the earth,[55] 3,000 to the Chinese border and 4,000 inside China itself.

Currently the pipeline has three lines in parallel, each running for 1,830 km. Construction of Line A/B commenced in July 2008. With pipe diameter of 1,067 mm, Line A became operational in December 2009, and Line B became operational in October, 2010. A delivery capacity of 30 bcm per annum was reached by the end of 2011. Construction of Line C was started in September 2012 with a designed capacity of 25 bcm per annum.[56] It became operational in June 2014. Turkmenistan will supply about 80 per cent of the enormous quantities to be sent through this pipeline.[57]

The most stunning aspect of the triple pipeline project is the urgency with which it was completed. In normal circumstances, a decade is the time frame for a pipeline to be conceived and delivered. A framework agreement on 3 April 2006 to the inauguration on 13 December 2009 is less than three years. It was a remarkable feat in that sense. The logistics and geography did not favour the project. It required a long pipeline across China to connect a Chinese port to the Central Asian onshore and offshore fields located on the western parts. A cross country line in Turkmenistan was also necessary to reach the border of Uzbekistan at Gedaim.

The implications of the project are not confined to the energy aspect alone. In many respects, it signifies a historic turning point in the geopolitics of the entire region, according to an insightful commentator of the area.[58] Turkmenistan could now feel a little less bound to the Russian route. It received a loan of $3 billion from Beijing for the development of the country's South Yolotan gas field. In return, Turkmenistan subsequently raised the amount of gas it committed to export to China through the pipeline from 30 bcm to 40 bcm and granted Beijing the rights to explore and develop the gas fields at South Yolotan to pay off the loan.[59] It was the first deal with a foreign company to develop an onshore energy field. The Turkmen government is open to foreign investment and ownership in oil and gas fields in the country's offshore section of the Caspian Sea,[60] but the onshore fields remain strictly with the domestic companies. The CNPC is the first of only two foreign companies – Eni of Italy later – that has the right to an onshore development, the Bagtyirlyk project near the Amu Darya River, through a 35-year production sharing agreement.[61] A consortium comprising the CNPC, the Petrofac Emirates from the UAE and LG International Corp and Hyundai both from South Korea was formed and won a $9.7 billion worth of contracts to drill and build gas plants in South Yolotan.[62]

On the flip side, Turkmenistan has had to borrow money from China to meet its share of the development costs, making it a debtor nation to China. That puts China in a much stronger position in negotiations on price and other issues.[63]

If Turkmens felt a little more confident regarding their export outlets, the Chinese, too, felt a little more confident regarding the source of their imports. Their quest for energy security is guided by their apprehension that the US could employ its naval superiority to impose a naval embargo against China in situations of conflict between the two.[64] The overland routes provide a guarantee of an alternative safe supply, if it came to that. Besides, China is forever looking for more and varied sources of energy, even in normal times.

It is the Gazprom that has lost out on more than one count. First, it stands in stark contrast to the Prikaspiiskii pipeline agreement negotiated by Russia with Kazakhstan and Turkmenistan in 2007. Whereas the Turkmenistan-China pipeline was negotiated, signed and built within three years, there has been little progress on the Prikaspiiskii pipeline.[65] Second, shortly after the pipeline to China opened, Gazprom had to negotiate an agreement with Turkmenistan to end their acrimonious dispute since the pipeline explosion in 2009. In late December 2009, Moscow reached an agreement with Ashgabat to buy 30 bcm annually of Turkmen gas starting in 2010 and to build a new pipeline to link untapped gas reserves in eastern Turkmenistan with the Prikaspiiskii pipeline.[66] Third, the choice of China has worked out well for the Central Asians. To that extent, Russia is on the back foot. China has tied in not just Turkmen, but also Kazakh, Uzbek and Tajik gas through the parallel lines. In future, there could be a scramble for the gas resources between Russia and China. China, in that scenario, is already there and has proven itself to be a speedy and reliable partner. The Russian sources of imports are shrinking in Central Asia and competition with China in future is possible. In the current situation, however, Russia has been forced to reduce the gas imports as it is coping with the sanctions, lower oil and gas prices and devalued rouble.

A short connection to Iran

Iran is one of the most hydro-carbon rich countries in the world. Like Turkmenistan, however, the energy geography in Iran is less than fortunate.[67] Its gas and oil fields are located in the southwest of the country and along the Gulf coast. The domestic market for energy is more or less concentrated in the north as the urban centres, industries and agro-industries are around the Caspian rim. Its population is growing rapidly, and the development of energy sector is lagging far behind the demand.

In 1997, a 200-km-long gas pipeline was inaugurated to carry natural gas from the Turkmen Korpeye gas field to Kurd Koi in northern Iran. The 2 bcm to be pumped annually was to be raised first to 4 bcm in 1999 and then to double that amount.[68] It was the first export pipeline from a CIS country after the break-up of the Soviet Union. Three years down the line, Iranian officials reported that Turkmenistan provided a total of only 6 bcm in the past three years, whereas the commitment was for 4 bcm a year. Iran complained that Turkmenistan never invested enough in the Korpeye field to make more gas available.[69] The poor performance sparked anger because

Turkmenistan was supposed to provide gas at no cost for the first three years to pay for the line that was built by Iran. Iran's annoyance seemed to have grown because terms of the contract allowed Turkmenistan to charge off its debt at a rate that was higher than its tariff for Russia and Ukraine. Under gas sales agreement, Turkmenistan was also committed to supply Iran at $40 for a thousand cm,[70] which was much lower than the price it could have expected.

The dispute erupted again in late 2007, when Turkmenistan halted the deliveries completely citing the disruption on technical problems. Iran linked it to a pricing dispute. Turkmenistan followed up the dispute by reaching an agreement with Gazprom to increase the price of gas by 30 per cent, which could be cited for a higher price from Iran. Iran sought a way out by slashing its daily sales to Turkey by around 75 per cent. At the peak of winter, in January 2008, it was forced to halt its gas exports completely to compensate for a domestic consumption crunch.[71] In early 2010 a new, second pipeline from Hangeran to Sangbast was launched, which was expected to deliver gas to Iran up to 20 bcm annually. A new gas-compressor station started operation in western Turkmenistan in December 2013, built specifically to export more gas to Iran.

In August 2014, there was yet another interesting development, when the Iranian Oil Minister Bijan Namdar Zanganeh said that his country no longer needed gas from Turkmenistan. 'Iran is importing Turkmen gas just because it is important to promote political and economic relations with Turkmenistan,' he added for an emphasis.[72] Whether the statement should be read as a ploy for better terms, or as Iran's expectations that the sanctions would soon be lifted and it would re-enter the gas market in a stronger position – only time will tell.

Recapitulation and leads

Turkmenistan is a landlocked country bounded by the Caspian Sea and the states of Afghanistan, Kazakhstan, Uzbekistan and Iran. The Caspian Sea is landlocked itself. Except for Iran, the rest of its neighbours are also landlocked. The Turkmen gas is considered high in hydrogen sulphide and carbon dioxide, and its quality is thought to be continuously deteriorating. The two factors together impinge on Turkmen gas production. In 2008, there was a sudden spurt in the estimates of its proven reserves after the survey done by the British firm Gaffney, Cline, and Associates (GCA), even though its findings are contested. President Gurbanguly Berdimuhamedov has renamed the oil and gas rich area Galkinish (meaning revival in Turkmen language), has designated the energy sector a priority sector and sought to develop and modernise it. The Turkmen gas policy rests on two pillars: one, it sells its gas at the border and takes no responsibility for the safe passage beyond it, and two, it permits only the domestic companies to develop onshore fields. An exception was made in the case of the CNPC of China and much later in the case of the Italian company Eni.

The Ukraine–Russia–Turkmen gas pipeline, till the break-up of the Soviet Union, was a domestic grid. The line has been a constant source of friction between Turkmenistan and Russia. In view of the situation in Ukraine, gas from Russia does not flow through the pipeline any more. The Central Asia–China Gas Pipeline starts at Turkmen-Uzbek border city Gedaim and runs through central Uzbekistan and southern Kazakhstan before reaching Horgos in China's Xinjiang Uygur Autonomous region. It is the longest line in the world stretching 7,000 km, which is over one-sixth the circumference of the earth, 3,000 km to the Chinese border and 4,000 km inside China itself. Turkmenistan will supply about 80 per cent of the enormous quantities to be sent through this pipeline. The pipeline was built within a very short period of three years. A short pipeline carries natural gas from the Turkmen Korpeye gas field to Kurd Koi in northern Iran. A second line was built later to enhance the delivery. Iran indicates that it may not need the Turkmen gas in future.

The TAPI pipeline will be studied in the next chapter. It is to be noted here that it is still in a preliminary stage. The Trans-Caspian Pipeline (TCP) trilateral project involving Turkmenistan, Azerbaijan and Turkey is not even on the horizon. In the present circumstances, Turkmenistan remains almost totally reliant on the Chinese route for the passage till the routes to Turkey and India open and perform. In view of the fact that the gas export is almost the only mainstay of its economy, Turkmenistan is in an unfavourable situation at present.

Notes

1 Anthony Rinna, 'Stepping Out of the Shadows: Turkmenistan and Its Feisty Neighbors', *Journal of Energy Studies*, 20 November 2013. http://ensec. org/index.php?option=com_content&view=article&id=471:stepping-out-of-the-shadows-turkmenistan-and-its-feisty-neighbors&catid=139:issue-content&itemid=425. Accessed on 30 February 2014.
2 Michel Chossudovsky, 'War Is Worth Waging: Afghanistan's Vast Reserves of Minerals and Natural Gas', *Oriental Review*, 19 June 2010. http:// orientalreview.org/2010/06/19/the-war-is-worth-waging-afghanistans-vast-reserves-of-minerals-and-natural-gas/. Accessed on 13 June 2011.
3 Hugh McDowell, 'Upstream and Downstream Oil and Gas Industry Potential in Turkey', *Caspian and Black Sea Oil and Gas Conference 2004*, Istanbul, 26–27 February 2004. Quoted in Michael Fredholm, 'Globalisation and Eurasia's Energy Sector', *The Journal of Central Asian Studies* (Srinagar), Vol. 20, no. 1, 2011, pp. 1–17.

See also Martha Brill Olcott, *The Geopolitics of Natural Gas: Turkmenistan: Real Energy Giant or Eternal Potential?* (Center for Energy Studies, Rice University's Baker Institute and Belfer Center for Science and International Affairs of the Harvard Kennedy School, 10 December 2013), p. 16.

4 'Turkmenistan: Bold New Production Strategy', *Petroleum Economist*, Vol. 70, no. 11, November 2003, p. 102. See also 'Turkmenistan Energy Profile: Some of World's Largest Gas Reserves – Analysis', *Eurasia Review*, 26 January 2012. http://www.eurasiareview.com/26012012-turkme nistan-energy-profile-some-of-worlds-largest-gas-reserves-analysis/. Accessed on 3 January 2013.

5 Maureen S. Crandall, *Energy, Economics and Politics in the Caspian Region: Dreams and Realites* (Praeger Security International, Westport, 2006), p. 101.

6 The Central Asia-Centre pipelines run from Turkmenistan via Uzbekistan and Kazakhstan to Russia.

7 Crandall, n. 5, p. 102.

8 Miyamoto Akira, *Natural Gas in Central Asia: Industries, Markets and Export Options of Kazakhstan, Turkmenistan and Uzbekistan* (The Royal Institute of International Affairs, London, 1997), p. 44.

9 Ibid, pp. 40–41. Akira considered the figures 'optimistic', referring to the controversy about the volume of unexplored reserves.

10 John C. K. Daly, *UPI*, 6 May 2008.

11 Quoted in Bruce Pannier, 'Independent Audit Shows Turkmen Gas Field "World Class"', *RFE/RL*, 14 October 2008. The Gazprom promptly dismissed the findings, as the earlier surveys done by the Soviet Union had come up with a much lower estimate. http://www.rferl.org/content/ Independent_Audit_Shows_Turkmen_Gas_Field_WorldClass_/1329822. html. Accessed on 30 November 2009.

12 After delivering a verdict of a low estimate of 4 tcm, a best estimate of 6 tcm and a high estimate of 14 tcm, Gillet seemed to have settled at his best estimate of 6 tcm.

13 'Turkmenistan's Yolotan-Osman Field Estimated a "Super Giant"', *Gas and Oil*, 20 February 2009. http://www.gasandoil.com/news/2009/03/ ntc91155. Accessed on 30 April 2012.

14 Dmitry Solovyov, 'Turkmen Gas Field to Be World's Second-Largest', *Reuters*, 25 May 2011. There are claims that 99.5% of Turkmen territory is conducive to prospecting. Abdelghani Henni, 'Gas for Cash: The Future of Turkmenistan', *Society of Petroleum Engineers*, 11 November 2014. http:// www.spe.org/news/article/Turkmenistan-Gas-for-Cash. Accessed on 21 November 2014.

15 John Foster, 'Afghanistan, the TAPI Pipeline and Energy Geopolitics', *Journal of Energy Security*, 23 March 2010. http://www.ensec.org/index. php?option=com_content&view=article&id=233:afghanistan-the-tapi- pipeline-and-energy-geopolitics&catid=103:energysecurityissuecontent&I temid=358. Accessed on 14 June 2012.

16 *Turkmenistan.ru*, 20 November 2011. http://www.turkmenistan.ru/en/ articles/15619.html. Accessed on 30 December 2013.

17 Putin proposed forming a Eurasian natural gas alliance during a meeting with Niyazov on 21 January. 'The alliance would make it possible to exert effective control over the volumes and directions of Central Asia's gas

exports, would create a unified balance of production and consumption of gas, and would ensure its export through a single export channel,' Putin said. He added that the alliance would 'bring an element of stability into the transportation of gas on a long-term basis'. http://www.thefreelibrary. com/Putin+suggests+Eurasian+natural+gas+alliance.+ (International . . . - a084548476. Accessed on 30 August 2013.

18 12 July 2002. http://www.turkmenistan.ru/?page_id=6&lang_id=en& elem_id=5662&type=event. Accessed on 13 November 2012.
19 Marat Gurt, 'Turkmens Seek U.N. Support for Afghan Gas Project', *Reuters*, 24 April 2002.
20 The CIS was founded in 1991 as a successor entity to the Soviet Union, retaining sovereignty of member states. Its members are Russia, Belarus, Armenia, Azerbaijan, Kazakhstan, Kyrgyzstan, Moldova and Uzbekistan. Turkmenistan and Ukraine have chosen to downgrade their memberships to the associate memberships. Georgia has withdrawn from the group.
21 ECO was established in 1985 by Iran, Pakistan and Turkey as an economic, cultural and technical cooperation group. In 1992, it was expanded to take in Afghanistan, Azerbaijan, Kazakhstan, Kyrgyzstan, Tajikistan, Turkmenistan and Uzbekistan.
22 *IRNA*, 3 December 2000.
23 Akira, n. 8, p. 48.
24 'Geologists Prepare the Discoveries', *Turkmenistan: The Golden Age*, 20 April 2015. http://turkmenistan.gov.tm/_eng/?id=4701. Accessed on 23 May 2015.
25 Under Berdimuhamedov, Turkmenistan has moved away from the era of 'Sultanism' under Niyazov to the one of 'neopatrimonialism' with an entrenchment of localised network of elites based on kinship, according to an analysis. Nicholas Kunysz, 'From Sultanism to Neopatrimonialism? Regionalism within Turkmenistan', *Central Asian Survey*, Vol. 31, no. 1, 2012, pp. 1–16.
    A totally different perspective holds that all of the conditions that led to Niyazov's sultanistic regime were still in place: a small, impoverished population, state control over all major industries and massive reserves of natural gas that free the government from dependence on a social base for support. S. Mitas, 'Turkmenistan: Bad Times Never Seemed So Good', *Central and Eastern European Online Library Transitions Online*, 2 October 2009. http://www.ceeol.com. Accessed on 13 November 2013.
26 Quoted in John C. K. Daly, 'TAP Pipeline Reality or Romance?', *UPI*, 6 May 2008.
27 'Turkmenistan Cares for Its Reliable Energy Resources Reputation', *Gas and Oil*, 14 April 2009. http://www.gasandoil.com/news/2009/05/ ntc92179. Accessed on 30 January 2014.
28 Solovyov, n 14. Also see 'Turkmenistan Energy Profile: Some of World's Largest Gas Reserves – Analysis', n. 4.
29 Devlet Atabayev, head of the London office. Quoted in 'Turkmenistan's Yolotan-Osman Field Estimated 'Super Giant''', *Gas and Oil*, 20 February 2009.http://www.gasandoil.com/news/2009/03/ntc91155.Accessedon 13 January 2013.
30 On 18 November 2014, Eni signed an agreement to extend the Production Sharing Agreement (PSA) till 2032 on the onshore Nebit-Dag.

18   November   2014.   http://www.eni.com/en_IT/media/press-releases/2014/11/Eni_strengthens_its_presence_Turkmenistan_Des calzi_Berdimuhamedov_Begdjanov_Kakayev.shtml.   Accessed   on   11 December 2014.

31 The country welcomes foreign participation in its offshore energy projects, but the onshore ones are reserved for the domestic companies. 'Turkmenistan's New Energy Policy Shows a Clear Preference for China and the West', *ISN Security Watch*, 24 August 2010. http://oilprice.com/Energy/Natural-Gas/Turkmenistans-New-Energy-Policy-Shows-A-Clear-Preference-For-China-And-The-West-.html. Accessed on 30 January 2014.

32 John Foster, n. 15.

33 Rinna, n. 1.

34 In 1994, NATO offered the Euro-Atlantic countries to associate themselves with the NATO in every field of NATO activity, including defence-related work, defence reform, defence policy and planning, civil-military relations, education and training, military-to-military cooperation and exercises, civil emergency planning and disaster response and cooperation on science and environmental issues. There are 22 members at present. http://www.nato.int/cps/en/natolive/topics_50349.htm. Last updated on 31 March 2014. Accessed on 23 June 2014.

35 Akira, n. 8, pp. xvii–xviii.

36 Ibid, pp. 50–51.

37 Sudha Mahalingam, 'India-Central Asia Energy Cooperation', in Santhanam K. and Ramakant Dwivedi, eds., *India and Central Asia: Advancing the Common Interest* (IDSA and Anamaya Publishers, New Delhi, 2004), p. 122.

38 Also see Chapter 4.

39 Stephen Blank. Quoted in N. J. Watson, 'Russia Secures Turkmenistan Gas Exports Until 2028', *Petroleum Economist*, Vol. 71, no. 10, October 2004, pp. 22–23.

40 Isabel Gorst, 'Broadening Export Strategy', *Petroleum Economist*, Vol. 71, no. 5, May 2004, p. 21.

41 Simon Tisdall, 'Cheap Gas or Democracy? US's Turkmen Problem', *The Guardian*, 26 January 2007.

42 The Pre-Caspian pipeline was to be routed via Turkmenistan for about 360 kilometres and another 150 kilometres via Kazakhstan before connecting with the existing Central Asia–Centre gas pipeline network on the Russia–Kazakhstan border. Also called the Central Asia–Centre line, the pipeline is the key route through which Turkmenistan exports its gas to the Gazprom's natural gas system. The western branch delivers Turkmen natural gas from near the Caspian Sea region to the north, while the eastern branch pipes natural gas from eastern Turkmenistan and southern Uzbekistan to western Kazakhstan.

43 Lili di Puppo, 'What Future Is There for the Trans-Caspian Pipeline after the Turkmenbashi Deal?', *Source: Caucaz.com*, 23 June 2007. http://www.gasandoil.com/news/2007/07/ntr72897. Also, Daniel Sershen, 'Turkmenistan's Natural Gas: Mixed Blessing', *The Christian Science Monitor*, 15 May 2007.

44 Robert M. Cutler, 'Turkmenistan Confirms Export Shift Away from Russia', *Eurasian Security Foresight Briefings*, 1 September 2010. http://www.

eurasiansecurity.com/research-analysis/foresight-briefing/turkmenistan-confirms-export-shift-away-from-russia. Accessed on 13 September 2013.

45 Alexandos Petersen, 'Pop Goes the Pipeline', *Foreign Policy*, 20 May 2009. http://foreignpolicy.com/2009/05/20/pop-goes-the-pipeline/. Accessed on 14 December 2014.

46 Quoted in Bruce Pannier, 'Pipeline Explosion Raises Tensions between Turkmenistan, Russia', *RFE/RL*, 14 April 2009. Also, 'Turkmenistan Cares for Its Reliable Energy Resources Reputation', *Gas and Oil*, 14 April 2009. http://www.gasandoil.com/news/2009/05/ntc92179. Accessed on 20 December 2014.

47 Vladimir Milov, 'Ups and Downs of the Russia–Turkmenistan Relationship', in Adrian Dellecker and Thomas Gomart, eds., *Russian Energy Security and Foreign Policy* (Routledge, New York, 2011), p. 16.

48 Petersen, n. 45.

49 Catherine A. Fitzpatrick, 'Turkmenistan: Chinese Deal Helps Stall Trans-Caspian Pipeline, Deter Caspian Conflict', *Eurasianet*, 30 November 2014. http://www.eurasianet.org/node/64610. Accessed on 5 January 2016.

50 Catherine Putz, 'Turkmenistan Completes East-West Pipeline: What's Next?', *The Diplomat*, 29 December 2015. http://thediplomat.com/2015/12/turkmenistan-completes-east-west-pipeline-whats-next/. Accessed on 2 January 2016.

51 Foster, n. 15.

52 Daniel Kimmage, 'Turkmenistan–China Pipeline Project Has Far-Reaching Implications', *RFE/RL*, 10 April 2006.

53 Bruce Pannier, 'Gas-rich Turkmens Shiver in the Cold', *RFE/RL*, 3 January 2008.

54 'Construction on Third Line Begins for Central Asia-China Gas Pipeline', *Pipeline International*, March 2012. http://pipelinesinternational.com/news/construction_on_third_line_begins_for_central_asia-china_gas_pipeline/066998/. Accessed on 31 January 2014.

55 Felix K. Chang, 'Friends in Need: Geopolitics of China–Russia Energy Relations', *Foreign Policy Research Institute*, May 2014. http://www.fpri.org/articles/2014/05/friends-need-geopolitics-china-russia-energy-relations. Accessed on 3 August 2014.

56 *China National Petroleum Corporation*. http://www.cnpc.com.cn/en/FlowofnaturalgasfromCentralAsia/FlowofnaturalgasfromCentralAsia2.shtml. Accessed on 13 May 2015.

57 Ole Odgaard and Jorgen Delman, 'China's Energy Security and Its Challenges towards 2035', *Energy Policy* (Amsterdam), Vol. 71, 2014, p. 111.

58 M. K. Bhadrakumar, 'China Dips Its Toe in the Black Sea', *Asia Times*, 1 August 2009.

59 *Central Asia Caucasus Analyst*, 20 January 2009.

60 For example with Russia's Itera and Germany's RWE Dea.

61 'Turkmenistan Energy Profile: Some of World's Largest Gas Reserves – Analysis', n. 4.

62 Solovyov, n. 14.

63 Abdujalil Abdurasulov, *BBC News*, 20 November 2014.

64 Marat Gurt, 'China Asserts Clout in Central Asia with Huge Turkmen Gas Project', *Reuters*, 4 September 2013.

65 Stephen Blank, 'The Strategic Implications of the Turkmenistan-China Pipeline Project', *Gas and Oil*, 16 February 2010. http://www.gasandoil. com/news/2010/04/cnc101536. Accessed on 13 March 2013.

66 *Financial Times*, 22 December 2009.

67 Also see Chapter 3.

68 Dilip Hiro, 'Caspian Gas: New Pipeline Benefits Iran', *Middle East International*, no. 568, 13 February 1998, p. 16.

69 Michael Lelyveld, 'Turkmenistan: Iran Boosts Gas Imports', *RFE/RL*, 27 February 2001.

70 'Iran Seeking Increase in Turkmen Gas Imports – IRNA', Quoted in *Emerging Markets Report* (Dow Jones & Company), 25 February 2001.

71 'Iran Expects Gas Flap with Turkmenistan to End This Week', *Daily Star*, 3 January 2008.

72 Qishloq Ovozi, 'Is Turkmenistan Losing Iran as a Customer?', 14 August 2014. http://www.rferl.org/content/qishloq-ovozi-turkmenistan-iran-gas/ 26530894.html. Accessed on 2 January 2016.

Part III

THE HOME TRUTHS

6

INDIA

Not a single transnational pipeline yet

Natural gas is emerging as the preferred fuel of the future in view of it being an environmentally friendly economically attractive fuel and also a desirable feedstock. Increased focus needs to be given to this potential sector.[1]

India is surrounded by a relatively gas rich neighbourhood. To the west of India the southern region of Iran has abundant reserves, for which the most logical and attractive market happens to be the Indian sub-continent. To the east Bangladesh is still finding huge quantities of gas which can't possibly be consumed in that country in the foreseeable future. Further east Myanmar is another potentiality rich source of supply. In addition there are of course possibilities of supply from other countries in the middle-east to the west of us; Central Asian republics to the north of us; and sources in south-east Asia to the east of us. With the prospect of high economic growth in India and the relative scarcity of indigenous supply of hydrocarbons, it would be of mutual advantage for India to pursue pipeline supply options from all feasible sources in our neighbourhood and arrive at contractual arrangements that through competition between these sources can prove to be of benefit to India.[2]

The quotes here endorse the general perception that gas is the preferred fuel of the future on account of it being economically attractive and environmentally friendly. They also offer Indian rationale and suggest the route to it: since India is surrounded by a gas-rich neighbourhood, it needs to acquire it from the neighbouring countries; since India is set on a high economic growth with a scarcity of indigenous supply of hydrocarbons, it needs to pursue pipeline supply option from all feasible sources, and it should opt for contractual arrangements to benefit from competition among the suppliers. The first quote comes from the *India Hydrocarbon Vision, 2025*. The then prime minister Atal Bihari Vajpayee placed it in both houses of parliament in March 2000. The second quote comes from the *Report of the Group on India Hydrocarbon Vision 2025*. After more than a decade and a half, not a single transnational pipeline has been constructed. Nor is an agreement on one finalised. The chapter seeks to inquire into the reasons thereof and to assess the losses therefrom.

Gas in/for India: an overview

History

The development of natural gas industry in India started in 1960s, when gas fields were discovered in Assam and Gujarat. The South basin fields were discovered in 1970s. More discoveries followed. The Krishna–Godavari (K-G) is considered the largest single gas find at 10–14 trillion cubic feet (tcf), that is 0.283 to 0.396 trillion cubic metres (tcm).[3] It was a Reliance discovery in 2002.

Initially, the exploration activities were carried out only by the national oil companies, like the Oil and Natural Gas Corporation Ltd (ONGC) and Oil India Ltd (OIL). It was the ONGC that discovered the South basin fields. Gradually, private companies were allowed to enter into exploration through joint ventures with the national oil companies. The New Exploration and Licensing Policy (NELP) of 1997–1998 provided equal platform to both public- and private-sector companies in exploration and production of hydrocarbons.[4] Today, the ONGC, the Gujarat State Petroleum Corporation (GSPC) and Gazprom are exploring the K-G basin; the British Petroleum owns part of the D6 field in the K-G basin and is developing the Panna–Mukta fields in the north-west of Mumbai on the western shore. The Royal Dutch Shell has invested in potential liquefied natural gas (LNG) facilities.

Reserves, demand, production, import

India's energy situation is at crossroads. It is characterised by increasing energy demand, high fossil fuel dependency, large import shares and a significant portion of population deprived of modern energy services.[5] India had 47 tcf (1.33 tcm) of proved natural gas reserves at the beginning of 2014. About 34 per cent of total reserves are located onshore, while 66 per cent are offshore.[6] The Indian gas reserves are estimated to constitute 0.5 per cent of the total world reserves. Since the countries bordering on the Indian shelf have made large oil and gas discoveries, there is an expectation that the Indian continental shelf deep water does contain recoverable hydrocarbons.[7] At the current production levels, the domestic reserves translate into an alarming reserve to production ratio of about 25 years against the world reserves and production ratio of about 61 years.[8]

There is a wide variation in the estimates of energy demand in India and its sector-wise composition. The Energy Information Administration (EIA) estimate puts the total energy demand in 2025 at 690 mtoe,[9] out of which the share of coal will be 43 per cent, oil and gas 52 per cent, nuclear 1 per cent and hydro 4 per cent. The updated EIA figures on India's gas production/consumption are as follows: India produces 1,551,303 terajoule (4 cm), it imports 725,463 terajoule (1.8 cm), the electricity plants use up 822,796 terajoule (2 cm), followed by industry, transport, residential and agriculture/forestry.[10] According to the British Petroleum (BP), there was a negative growth by 16.4 per cent in gas production in 2013.[11] It was the third consecutive year in decline. According to BP estimates till 2035, the demand for fossil fuel in India is set to expand, led by gas at 145 per cent, oil at 117 per cent and coal at 112 per cent.[12] Coal continues to dominate the energy consumption basket at 54 per cent followed by oil at 29.5 per cent, natural gas at 7.8 per cent and non-fossil fuels at 8.3 per cent.[13]

Even within the Government of India, there are different estimates of gas supply scenarios: the Ministry of Petroleum and Natural Gas puts the domestic production at 80.2 cm per day and imported re-gassified LNG at 41.11 cm per day, bringing the total to 121.13 cm per day.[14] As against this, the Hydrocarbon Vision estimates put the supply at 65 cm in 2000 and total demand for gas going up till 2020 to 391 cm per day.[15] The *Report of the Group on India Hydrocarbon Vision* (RGIHV, 2025) presents three production scenarios for 2020 – the optimistic one includes production from new fields and coal bed methane and puts the figure at 84 cm per day, whereas the pessimistic one puts it as low as 28 cm per day.[16] An independent study concludes that projections for the 2020 are particularly fraught with difficulty as realistic outcomes range from a huge demand increase to very little change.[17]

An interesting statistic on gas consumption in the country is an enormous fleet of cars run on the natural gas. Numbering 1,080,000 vehicles, India has the fifth-largest natural gas vehicle fleet in the world, behind Brazil, Iran, Argentina and Pakistan.[18] A steady increase in this number points to an increased demand but also to an improvement in climate condition.

Gas policies

The Hydrocarbon Vision 2025 makes two broad recommendations on the gas usage and procurement: one, timely and continuous review of gas demand and supply options to facilitate policy interventions; and two, pursuing diplomatic and political initiatives for import of gas from neighbouring and other countries with emphasis on transnational gas pipelines.[19] The RGIHV, 2025[20] translates the vision into policy prescriptions. It notes a fast-expanding energy demand and a declining production profile in the country. It takes note of the increasingly difficult areas in which the exploration and production would have to be carried out. Specifically, it mentions the deep waters where sub-sea pipelines would have to be laid at the water depths exceeding 3,000 metres to bring gas from West Asia. In the circumstances, it calls for alliances at the levels of the companies as well as the governments with a view to acquiring latest technologies and mobilising finances. The most significant prescription comes somewhere in the middle of the report on page 22: 'We have to forge linkages with large investors especially in Europe who can mobilise resources even into projects that attract US sanctions.' In 2000, the US sanctions were applicable to the energy-exporting countries like Iran, Iraq and Libya. The reference in the report seems to be Iran, with which the gas pipeline deal was going on for a decade and a half already by that time.

The Strategic Plan for 2011–2017 reiterates the problem of reaching farflung areas and difficult terrains in adverse climatic, logistic and security challenges, and then adds the problem of geopolitical unrests and instability as a further issue of concern.[21] This aspect of energy security/insecurity has steadily gone up in the country's policy options. In the late 1990s, there was regret among the energy experts that no comprehensive policy had been enunciated nor any 'white paper' published. Whereas the economic-related ministries, as well as those of External Affairs and Defence, were involved in separate and independent programmes to enhance energy security, there was no attempt to cooperate formally on this issue.[22]

An experts committee was appointed with the assignment to redress the deficiency in this respect. In its report,[23] the committee made specific recommendations for (1) energy efficiency in all sectors; (2) emphasis on mass transport; (3) active policy on renewable energy including bio-fuels and fuel plantations; (4) accelerated development of nuclear and hydro-electricity; (5) technology missions for clean coal technologies; (6) focussed research and development (R&D) on many climate-friendly technologies. The Petroleum and Natural Gas Regulatory Board would be given the mandate to moderate access to national gas pipeline network and ascertain transportation rates for common or contract carriers. On accessing gas from overseas, its prescription was to set up captive fertiliser and/or liquefaction facilities to augment energy availability. It was a fairly comprehensive report analysing each of the major energy sources, although it did not touch upon the issue of overseas gas pipelines in particular.

Structure

At the apex of the energy structure is the Ministry of Petroleum and Natural Gas (MoPNG). It is entrusted with the responsibility of exploration and production of oil and natural gas, their refining, distribution and marketing, import, export and conservation of petroleum products and the LNG.[24] Gas Authority of India Limited (GAIL) was incorporated as a public-sector undertaking (PSU) under the MoPNG in August 1984. It claims to be India's youngest *maharatna*[25] and has an increasingly global presence. Its wholly owned subsidiary company Gail Global (Singapore) pursues overseas business opportunities including LNG and petrochemical trading. Gail Global (US) has working interest in Carrizo Oil and Gas Inc., Eagle Ford Shale Acreage in Texas, and is exploring LNG-import/liquefaction-capacity booking opportunities from the US. It is an equity partner in Fayoum Gas Company and National Gas Company in Egypt. It is an equity partner in China Gas Holdings Limited and has formed a joint venture company – Gail China Gas Global Energy Holdings Limited for pursuing gas-sector opportunities primarily in China. It is part of a consortium in two offshore E and P blocks in Myanmar and holds participating interest in the South East Asia Gas Pipeline Company Limited Inc for transportation of gas produced from two blocks to China.[26]

In 2006 the government began to reform the gas pricing and created the Petroleum and Natural Gas Regulatory Board (PNGRB) to protect the interests of consumers and entities engaged in specific activities relating to petroleum, petroleum products and gas to promote competitive markets. To that end, the board has been mandated to regulate the refining, processing, storage, transportation, distribution, marketing and sale of petroleum, petroleum products and natural gas, excluding the production of crude oil and natural gas so as to ensure uninterrupted supply of petroleum, petroleum products and natural gas in all parts of the country.[27]

Oil India Limited (OIL) was incorporated in 1959 to expand and develop the newly discovered oil fields of Naharkatiya and Moran in the Indian North-East. Today, it claims to be the flagship national oil company with business interests straddling the entire hydrocarbon chain – from refining, pipeline transportation and marketing of petroleum products to exploration and production of crude oil and gas and marketing of natural gas and petrochemicals. It also holds through its wholly owned affiliate Indoil Montney Ltd, Canada, a 10 per cent interest in the LNG-destined natural gas reserves in northern British Columbia.[28]

OIL has over 100,000 sq. km of area for its exploration and production activities, most of it in the North-East, which accounts for its entire crude oil production and majority of gas production. Rajasthan is the other producing area of OIL, contributing 10 per cent of its total gas production. In addition, OIL's exploration activities are spread over onshore areas of Ganga Valley and Mahanadi. OIL also has participating interest in NELP exploration blocks in Mahanadi Offshore, Mumbai Deepwater, Krishna–Godavari Deepwater and so on, as well as various overseas projects in Libya, Gabon, the US, Nigeria and Sudan.[29]

In recent years, the OIL has increased its presence in the gas sector by significant investments in infrastructure such as LNG import terminal, gas pipeline and city gas distribution projects. When the GAIL was authorised to join the South Asia Gas Pipeline from Oman, the OIL made a strong case for joining the project as well, arguing that it 'needs to have substantial presence in the gas market to make up for the loss in liquid fuel business to provide total fuel solutions to the customers and to maintain its leadership in the oil and gas market'.[30] The pipeline would allow it to retain its relationship with liquid hydrocarbon customers and facilitate their switchover to natural gas, it wrote to the MoPNG.

New private companies such as Petronet LNG Limited have been formed in recent years to benefit from growing LNG imports in India by building re-gasification plants. It is a joint venture company prompted by the GAIL, ONGC, Indian Oil Corporation and Bharat Petroleum. It has set up the country's first LNG-receiving and re-gasification terminal at Dahej, Gujarat, and another terminal at Kochi, Kerala. While the Dahej terminal has a nominal capacity of 10 million metric tonne per annum (MMTPA) (equivalent to 40 cm of natural gas), the Kochi terminal has a capacity of 5 MMTPA (equivalent to 20 cm of natural gas). The company is in the process of building a third terminal at Gangavaram, Andhra Pradesh.[31] It has selected Gaz de France as its strategic partner. It has also signed an LNG sale and purchase agreement with Ras Laffan Liquefied Natural Gas Company Ltd, Qatar, for the supply of LNG to India.

Privately owned Reliance Industries (RIL) emerged as an important upstream player in the natural gas market after discovering significant reserves in the K-G basin in 2002. RIL started out as a small textile company to emerge as the largest private-sector enterprise and a Fortune 500 company.[32] The K-G basin gas production came online in 2009, but soon thereafter it got mired into a host of controversies on account of production, pricing and allocation.

The Gujarat State Petroleum Corporation Limited (GSPCL) is also developing several offshore areas in the K-G basin. Another promising producing area is the Cambay basin in western India, where independent company Oilex has done some preliminary work assessing the potential for 'tight' natural gas.[33]

No external pipelines

As the title of the chapter says, India has no external gas pipeline coming in, going out or traversing its territory. The domestic pipelines are far too inadequate. India has a unique distinction of having one of the lowest pipeline spreads per square kilometre of land at 0.003 kilometres (km) of pipeline compared to the United Kingdom with 1.08 and the US with 0.19 km of pipeline per square kilometres of land. India does not have a gas hub for transmission system, either.[34]

Till 2006, Gail functioned as a monopoly on operating gas pipelines within the country. Today, it operates the 14,987.6-km-long Hazirabad–Vijaipur–Jagdishpur–GREP–Dahej–Vijaipur (in short, HBJ Pipeline) line, which is the longest in the country. It plans to extend the same to the southern parts of India. Reliance Gas Transportation Infrastructure (RGTIL) owns and operates the East–West line from RIL's K-G basin and pumps it to India's northern and western markets. Petronet, Assam Gas Company and GSPC have plans to build their own lines to link the production sites to the network.

The pipeline construction has fallen far behind the set targets and suffers from several problems. Gail had to cancel the contract to build a 1,156 km-long Kochi–Kanjirkkod–Bangalore–Mangalore pipeline after insistent local protests. Similarly, the RIL authorisation to bring gas from the K-G basin was cancelled due to inordinate delays in its implementation.

There is also a problem of severe disparities in pipeline reach. The states like Gujarat, Maharashtra and the former Andhra Pradesh, which are closer to gas sources, have seen higher gas consumption, whereas Punjab, Haryana, Jharkhand, Karnataka, Kerala, Bihar and Chhattisgarh suffer as they have inadequate access to the pipelines. The tariff structure creates an additional hurdle to the pipeline construction. There are different tariffs to transport the same amount of gas across the same distance in different parts of the country.

The lack of pipelines impacts the growth of fuel directly. B. S. Negi, a former member of the PNGRB and a former director of planning at GAIL, considers the current tariff structure highly irrational. He points out that the 'infrastructure should precede the growth of supply, but that can happen only if the tariff structure is conducive to incentivise companies to build pipelines'.[35] The lack of pipelines also impacts the growth of LNG distribution directly. Till the pipelines are in place and the LNG terminals become operative, distribution of fuel, especially in the eastern and central parts of the country, would remain inadequate and the prospects of their economic development would remain less than satisfactory.

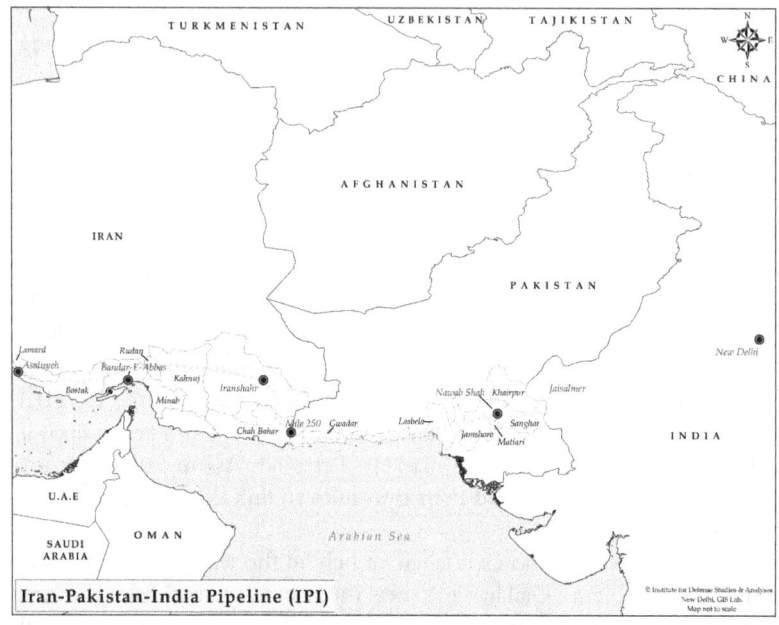

Figure 6.1 Proposed Iran–Pakistan–India pipeline.
Source: Institute for Defence Studies and Analyses.
Disclaimer: Map not to scale.

The RGIHV, 2025 recommends that the pipeline sector should be given the infrastructure status to treat it on par with other modes of transport.[36] The idea here is that if infrastructure status is given to LNG projects, only 4 per cent of Central Sales Tax will have to be paid as against the state sales tax rates, which vary widely from one state to another, and which would make LNG costlier than coal and jeopardise the investments being made by the companies in building LNG terminals.[37]

The then Indian oil minister S. Jaipal Reddy spoke at length on the issue at the 15th Foundation Day of the Petronet on 2 April 2012. 'We have a country-wide network of 12,000 km of gas pipeline (and another) 12,000 km of pipelines are under construction,' Reddy said. In addition, another 7,000 km are going through the bidding process, which should bring 30,000 km of pipelines by 2017, according to him. Relating the length of the pipeline to the passage of gas, he said that the present gas pipelines can transport around 8.1 billion cubic feet (0.23 bcm) of natural gas per day, but the added capacity could bring the country up to as much as 31 billion cubic feet (0.877 bcm) per day.[38] Today, the East–West Pipeline transports gas from Kakinada to Bharuch, and the North–South pipeline transports gas from Dabhol to Bangalore, in addition to the HBJ pipeline. In August 2014, Prime Minister Narendra Mody announced the establishment of a 2,050-km-long natural gas 'Energy Highway' connecting Jagdishpur–Phulpur–Haldia pipeline linking with the national gas grid.[39] The line will serve Bengal, Bihar, Jharkhand and Utter Pradesh.

Gas pipelines

Iran–Pakistan–India pipeline

The Iran–Pakistan–India (IPI) gas pipeline project has been under consideration for nearly two decades. Its origins can be traced back to 1998, when the Iranian Mostazafan va Janbazan Foundation announced its plan to cooperate with Shell, British Gas and Petronas to lay a 1,400-km gas pipeline between Iran's South Pars gas field to the port city of Karachi in Pakistan. Next year, India signed a preliminary agreement with Iran on the project. There are three possible routes for the proposed project: one, overland, from Iranian gas fields terminal at Assaluyeh to the Pakistani border on to India (the above map shows the overland route); two, offshore, hugging the Pakistani coast along Baluchistan; and three, deep sea, under the Pakistani exclusive economic zone (EEZ). The overland option would involve 56-inch-diameter pipeline stretching 2,000 km: 850 km in Iran, 700 km in Pakistan and 1,120 km in India with initial estimate of $700 million.[40]

The Indian defence experts reject the option of an overland route through Pakistan, as it empowers Pakistan to cut off supplies to India whenever it wants. International guarantees can reduce, but not eliminate, the chances of such a cut-off.[41] Pakistan favours an overland route as this will assure its own gas supplies while giving it a stranglehold over supplies to India. India favours the laying of the pipeline through the Pakistani EEZ. Such a pipeline will need to be sunk 3,000 metres and would be the first experiment in pipeline construction at such depth. Till recently, Pakistan declined to permit it, citing various techno-economic, defence and strategic reasons. It maintained that the project would cost one and a half time more if laid under water, with an estimated cost touching $10 to 12 billion. There were talks of involving stakeholders like the World Bank, the Asian Development Bank, international financial institutions and the private sector in both India and Pakistan.

During the Indian foreign minister Jaswant Singh's visit to Iran in July 2000, a joint working committee was set up to examine the security, political, technical and economic issues related to the transportation of the Iranian gas. In November, the Iranian deputy foreign minister S.M.H. Adeli visited India with a fresh proposal in this regard. According to it, India and Iran would consider the project 'bilateral'; none of them would need to enter into any contractual obligation with Pakistan or rely on its security assurances. That responsibility would lie with the international consortium that would run the project and manage the risks of disruption involved.[43] In 2002, the Gazprom announced its plan to construct an undersea pipeline at the depth of 150 metres through Pakistan's territorial waters. It was a significant step towards the realisation of the project.[44]

After the change of government in India in mid-2004, the then foreign minister K. Natwar Singh expressed India's willingness to consider the overland route, if Pakistan was willing to give required guarantees. Interestingly, his statement did not reiterate the Indian condition that Pakistan should confer the status of the Most Favoured Nation on us. The Pakistani response, predictably, was prompt and positive.[45]

In early 2005, there were positive statements as the Indian government agreed to give it a fresh look. Speaking in Islamabad on 16 February, the Indian foreign secretary Shyam Saran said at a press conference that India had agreed to go ahead with the project and that three countries were moving forward in a pragmatic fashion to make it a reality. He also spoke of an overarching trilateral agreement on the issue.[46] The statement was significant in view of the fact that India, till then, was in favour of separate agreements with Iran and Pakistan.

In the meanwhile, there were major deals on oil and gas between India and Iran. The two countries signed a preliminary agreement estimated at $40 billion that committed India to import LNG and develop Iranian oil fields and a gas field. The ONGC Videsh Limited (OVL) would help in developing the oil fields in exchange for 90,000 barrels a day (bd) of crude. India would also import 7.5 million tonnes of LNG, starting 2009 for a period of 25 years.[47] India would pay $1.2 plus 0.065 of Brent crude average, with an upper ceiling of $31 per barrel of LNG. Iran would ship 5 million tonnes of LNG annually with a provision to increase the quantity to 7.5 million. China would retain its 50 per cent share in Yadavaran, Iran's share would go down to 20 per cent and India would acquire the rest of the 30 per cent. India also acquired 100 per cent right in the Juffair oil field that is estimated to yield 300,000 bd.[48]

It is necessary, at this stage, to put the trilateral project into a broader context. The US considers Iran a hostile state and enforces unilateral economic sanctions against it. The US companies cannot enter Iranian gas market under the Executive Order 12959, which was signed by President Bill Clinton in 1995 and renewed by President Bush in March 2004. Notwithstanding comprehensive unilateral sanctions against Iran, the Iran and Libya Sanctions Act (ILSA) was enacted by Congress in August 1996.

ILSA had many of the same objectives as the unilateral sanctions but is different in jurisdictional scope. Unlike the embargoes against Iran and Libya, which were primary sanctions, ILSA imposed a secondary boycott. The legislation was designed essentially to force foreign companies into choosing to do business with Iran and Libya or the US.

ILSA mandated the US president to impose sanctions on any US or foreign person, company or country who, after 5 August 1997, invested $20 million or more in an Iranian project ($40 million for Libya; this was lowered to $20 million in August 2001 and has now been lifted), if the investment directly and significantly contributed to the enhancement of Iran's (or Libya's) ability to explore for, extract, refine, or transport by pipeline its oil and natural gas reserves. ILSA required that sanctions be imposed for a minimum of two years.

In June 2004, the US president George W. Bush and the Indian prime minister Atal Bihari Vajpayee signed the 'Next Step for Strategic Partnership' after a series of talks between the two countries. Several more agreements followed tying the two into a global partnership. 'New Framework for the US–India Defense Relationship' created and institutionalised a Defence Policy Group consisting of the senior leadership of the defence establishments of the two countries in June 2005.

During Prime Minister Manmohan Singh's visit to Washington in July 2005, a historic 'India–US Joint Statement' was signed. It resolved to establish a US–India 'global partnership' through increased cooperation on economic issues, on energy and environment, on democracy and development, on non-proliferation and security and on high-technology and space. On 2 March 2006, the US president and the Indian prime minister signed the 'Civil Nuclear Co-operation Agreement' during the former's visit to India. The evolving nuclear deal between India and the US impacted India's foreign policy, particularly with regard to Iran. India repeatedly cast votes against Iran on the nuclear issue indicating a rethink in Indian policies vis-à-vis Iran, which impacted the issue of the IPI.

There was a parallel track in India regarding gas imports around that time. It was almost as if there was no communication between those who cherished the reach to the US and those who promoted the pipelines across Asia. The then minister of petroleum and natural gas Mani Shankar Aiyar proposed a pan-Asian gas grid. The IPI line, under this scheme, could be extended to Burma and South China; Caspian gas could be delivered to Lebanon and Egypt through a series of pipelines connecting to the Blue Stream. For India, he pitched for all the three proposed pipeline projects – Indo-Iran, Indo-Myanmar and Turkmenistan–Afghanistan.[49] A few months later, Aiyar again expressed India's willingness to go ahead with the IPI project. He said he sought improvement in the World Trade Organisation

(WTO) rules so that pipelines are explicitly covered under Article 5 of the General Agreement of Trade and Transit (GATT)[50] dealing with freedom of transit. Originally introduced for ensuring trade channels of landlocked countries, the 'WTO Security Shield' would help India make sure Pakistan does not disrupt flow of gas.[51]

In between the two major initiatives, there was an extremely uncharacteristic outburst of anger and frustration from Aiyar. Unfortunately, it came on the day when the US secretary of state Condoleezza Rice was in Delhi to dissuade India from joining the IPI. Either as a consequence thereof or in an unfortunate coincidence, Aiyar called on Iran to set a reasonable price for its gas sale to India and Pakistan in view of the fact that the consumer countries constituted an attractive market of about 200 cm a day. 'Otherwise we would be forced to look at other options,' he said. In an unseemly taunt, obliquely referring to the US sanctions, he added that if Iran had alternative markets in the US, China and Japan, then these could be tapped by it.[52]

Speaking in New Delhi, Rice said that the US had conveyed its 'concerns' to India on the gas pipeline. 'Our ambassador [to India] has made statements in that regard. So, those concerns are well known to India.' As an incentive to wean India away from the project, she said, '[W]e believe that a broad energy dialogue should be launched with India.'

Two days later, Rice reiterated her stand in Islamabad. The US had imposed sanctions against Iran for 'good reasons', she said and offered to pool technologies and thinking to 'meet the burgeoning demand for energy'.[53] Ominously, she declined to speculate on the US response if India and Pakistan went ahead with the project.[54] Within days thereafter, the Pakistani prime minister Shaukat Aziz made a presentation in Islamabad where he said that the country was studying various options to acquire gas including the LNG from Iran, Qatar or Turkmenistan. Pakistan would make a final decision keeping in view the 'economic feasibility and the national interest'. His remarks were understood to have been made within the context of the controversy around the Iran pipeline; although he did not mention it in his presentation.[55]

The US made several diplomatic moves on the issue in quick succession thereafter. The US under-secretary of state Nicholas Burns wrote an article in *Washington Post* encouraging New Delhi to establish energy ties with Central Asian states and to reduce 'the lure of long-term contracts with Iran'. Talking to Indian TV channels on phone from Washington, he said, 'We hope that India as well as all other states – China, Russia, France, Britain and Japan – will diminish their economic relations with Iran.' He said he expected India to be 'part of the international mainstream in trying to deal with one of the most difficult security problems we face internationally today'.[56] On a visit to India, Henry Kissinger, the former US secretary of state, asked Murli Deora, the newly appointed minister of petroleum and natural gas,[57] not to have energy ties with Iran.[58]

The US energy secretary Samuel Bodman queried his Indian counterpart Deora regarding the pipeline during his visit to India, declaring that Washington needed to 'stop' it. The chairman of the Foreign Affairs Committee of the US House of Representatives, Sam Lantos, led a group of US Congressmen in writing a letter to the Indian prime minister suggesting that India pull back not just from the pipeline but also from the LNG deal with Iran. He introduced a bill in the US Congress entitled 'Iran Counter-Proliferation Act'. It was meant to adversely impact the already poor prospects of the pipeline. And to cap it all, the then assistant secretary of state for Non-proliferation and International Security, Stephen G. Rademaker, stated in New Delhi in February 2006 that India's votes in the International Atomic Energy Agency against Iranian nuclear programme were 'coerced'.[59]

In April 2008, the Iranian president Mahmoud Ahmadinejad visited New Delhi and Islamabad to garner support for the project. After meeting the Indian prime minister Manmohan Singh, Ahmadinejad described the discussions as positive and hoped that the deal would be finalised in the 'near future'. Foreign Secretary Shiv Shankar Menon reaffirmed Ahmadinejad's optimism but cautioned that a long road lay ahead to ensure that the project was commercially viable and financially acceptable to India and security concerns were taken care of. 'We think it is doable,' he said.[60]

India had stopped attending trilateral meetings on the pipeline for an entire year after the talks had broken down over transit fee issue between India and Pakistan and over pricing issue between India and Iran. Fallout of Ahmadinejad's visit was that the Indian officials agreed to resume attending the trilateral meetings.[61] Deora made positive comments during his visit to Islamabad saying that India wanted to activate both the IPI and Turkmenistan–Afghanistan–Pakistan–India (TAPI) pipelines, as both were important to India. Returning from Pakistan, he said that India and Pakistan had almost worked out a general agreement on the 'transit fee'.[62] M. K. Narayanan, the national security advisor, also chipped in with his views on India's civilisational and economic ties with Iran, referring to India's Shia population and cautioning that 'any mishandling will impact negatively on us'.[63]

Security implications for India

Ahmadinejad's visit to Islamabad yielded an addition to the prospective list of participants – China. A foreign ministry release in Islamabad said that

President Parvez Musharraf and Ahmadinejad had discussed the inclusion of China and the latter had welcomed it. The Pakistani foreign minister Shan Mahmood Qureshi added that the pipeline to China would be along the Karakoram Highway.[64] That probably has put a stop to any further talk on the issue as far as India is concerned – although not necessarily so.

The Indian strategic community and also the defence establishment have consistently opposed the project from the security angle.[65] The area on the Iran–Pakistan border is earthquake prone. There is an ongoing insurgency among the Baluch people both in Pakistan and in Iran. The Sindh Province in Pakistan is also disturbed. In such circumstances, gas pipeline is bound to be targeted.[66] An informed commentator conjures up yet another frightening scenario: taking a cue from the Ukrainian situation, India must also take on board a contingency that there could be a dispute between Pakistan and Iran, disrupting the supply. Since Pakistan is under the US influence and Iran is under the US sanctions, such a contingency is not too far-fetched.[67] Finally, the pipeline is scheduled to deliver only 30 cm per day, which is not very substantial and not worth the efforts and expenses to be incurred.

Price considerations for India

A well-researched cost calculation concludes that the cost to India would work out to be far more than optimal. It examines two measures that are normally used to arrive at the optimum price. The first is a cost-plus mechanism, which assumes a fair rate of return to the provider. The second is to consider alternative fuels for the end-user to calculate a competitive cost. On both of these measures, IPI does not pass the economic criterion. On the first measure, the transportation from the Indian border to a point where it could connect to the Hazira–Bijaipur–Jagdishpur pipeline would be enormous. On the second measure, higher-cost gas is justifiable only for certain industries such as fertilisers and petrochemicals, but their gas needs are not large when compared to the power sector. The alternative fuel option, the coal, is a cheaper and better answer to the power needs.[68]

A stalled initiative for the Iranian export of LNG has not helped with the IPI either. A 25-year, $22 billion agreement signed in 2005 requires India to build an LNG plant in Iran. The plant would need American components, which might violate the US sanctions. The price for the LNG imports has also run into trouble and been added to the price dispute over gas for the IPI. The national oil companies of Iran and India disagree about the legal interpretation of the contract for the export of LNG to India. This deal was signed in 2005 before Mahmoud Ahmadinejad was elected president of Iran and was tied to a relatively low market price for crude oil.

India considers the deal final and binding, while Iran has argued that it is not binding because it has not been ratified.[69]

New Delhi's final position rests on three conditions even as it remains interested in the pipeline project: it would pay for the gas only after it is received at the Pakistan–India border, it would not pay penalty in case of a delay and it is opposed to Iran's demand to revise the deal's gas prices every three years.[70] As the process of lifting the sanctions is already under way, the situation at the Iran end is about to undergo a drastic change.

Future of sanctions

There is a clear understanding of the changing circumstances in Iran among the energy observers in India, who have drawn attention to this fact. The Associated Chambers of Commerce and Industry of India (ASSOCHAM) has put out a study advocating that 'India must fully exploit the economic opportunities unfolding from the lifting of Western sanctions on Iran', even though its preference remains a route bypassing Pakistan.[71] A policy brief from the Institute of Defence Studies and Analysis (IDSA) also recommends an early revival of energy ties with Iran, as once the sanctions are lifted, Iran may prefer to send its gas to the more lucrative European market.[72] M. K. Bhadrakumar, a knowledgeable commentator on the issue, notices triggering reverberations in regional politics, as Tehran, Islamabad and Beijing are positioning themselves to tap into the new vistas opened by Iran's integration into the international community.[73]

The Iranians have been quietly confident of the IPI winning the pipeline war all through the tortuous process.

Turkmenistan-Afghanistan-Pakistan-India Pipeline

From its very inception, TAPI was a competitor to the IPI. In fact, it was meant to replace the IPI. A dramatic illustration of the fierce rivalry between the two was the visit of the Turkmenistan president Gurbanguly Berdimuhamedov to Kabul and the Iranian president Ahmadinejad to Islamabad, both on the same day, that is 28 April 2008. Both the presidents were on their respective missions to enlist the transit countries – Afghanistan and Pakistan – into their respective pipeline plans.

The origins of TAPI go back further than the IPI. In 1995, the Centgas consortium initiated by the US oil company Unocal – in collaboration with Saudi Arabia's Delta Oil, Argentina's Bridas, Turkmenistan's Turkmenrosgaz, Russia's Gazprom and others – proposed and aggressively pursued an energy route. It envisaged a double pipeline, an oil pipeline that could transport some 50 million tonnes of oil per year from the Turkmen oil field

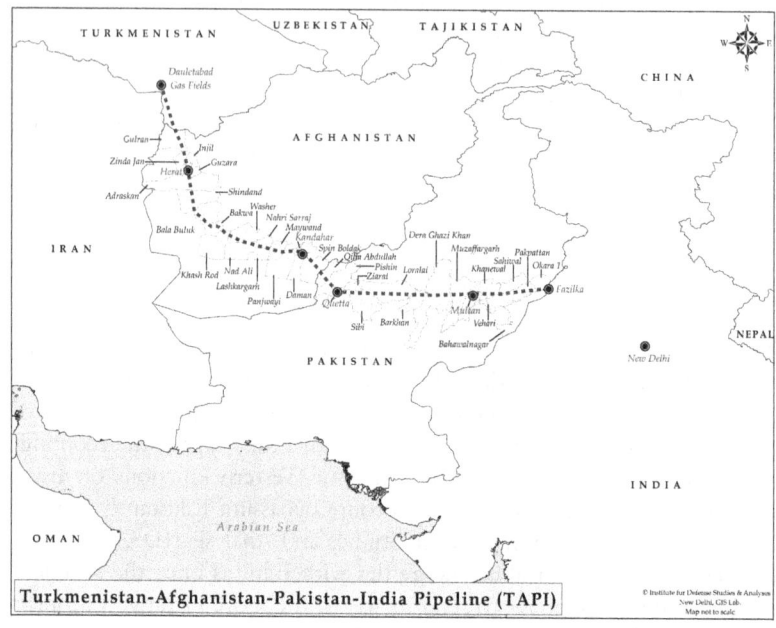

Figure 6.2 Turkmenistan–Afghanistan–Pakistan–India pipeline: TAPI route.
Source: Institute for Defence Studies and Analyses.
Disclaimer: Map not to scale.

of Chardzou to a terminal located on Pakistan's Indian Ocean coast. From
there the oil could be exported to the rest of the world including India.
A gas pipeline could transport some 20 bcm annually from the Dauletabad gas field in southeast Turkmenistan to Multan in Pakistan through
Herat and Kandahar in Afghanistan.[74] In Afghanistan, it would skirt around
the mountains in the west and south reaching Quetta before arriving at
the Indo-Pakistan border near Pakpattan to avoid the insurgency-infested
Balochistan and eastern Afghanistan. On the initial estimate, the project
would have required $2.5 billion external financing. The ADB completed
a feasibility study in 2005 that was updated in 2008. Details were outlined
at the April 2008 meeting of the four participating countries. The ADB
reported that the estimated capital cost was $7.6 billion and said it would
consider financing for the project.

The plan suffered from two serious problems. First, the situation in
Afghanistan ruled out safe passage of the line. The international lending
agencies determined that the situation in Afghanistan precluded lending

until such time as a stable regime had been established.[75] Insecurity could range from the kidnapping of the expatriate staff, to financial blackmailing, to blowing up pipelines.[76] Extortion and protection money posed serious risks. The participants were not deterred on this count. The optimism of Turkmenistan and the energy companies was rooted in their cynical realisation that the Afghan warlords needed the funds generated by the pipeline for the purchase of armaments.[77] After lengthy negotiations, in January 1998, Taliban formally agreed to the pipeline project and pledged to provide full security for it.

The second problem lay within Turkmenistan itself. The country had made export commitments that were not seen to be feasible. Just as the talks on the TAPI pipeline progressed, so did the Turkmen agreements with its customers. In 2008, Turkmenistan agreed to extend its export of 50 bcm to Gazprom by two more decades. The export to China for 30 bcm a year was to begin the next year. A deal with Iran committed 8 bcm annually. And an agreement with the EU promised to send 10 bcm per year. The acceptance of the TAPI pipeline would have brought the export figure to 100 bcm.[78] Even the newly discovered gas deposits were not seen to be adequate to yield such enormous amounts of gas. An Asian Development Bank–commissioned survey cast doubt on the field's potential to feed the TAPI pipeline for 25–30 years, a time frame necessary to make the project viable. In late 2009, Turkmen officials stated they would offer gas from the Yasrak field instead of the planned Dauletabad field. They provided a reserves certification for Yasrak.[79]

The TAPI pipeline had a strong advantage on the other hand: the US support. It answered many of the American preferences: it kept Iran and Russia out of the global energy map; it could wean away both Pakistan and India from the IPI; it helped Turkmenistan to diversify its gas exports away from Russia. And finally, it furthered the American Silk Road project connecting Central Asia to Afghanistan. The location of Afghanistan was indispensable in that architecture. The country is ultra-strategic: positioned between West Asia, Central Asia and South Asia, between Turkmenistan and the avid markets of the Indian subcontinent, China and Japan. Afghanistan is at the core of 'Pipelineistan'.[80] A revisionist narrative of the US war on Afghanistan goes like this: 'Afghanistan is an invaluable prize, and the US went to war on the country in 2001 to keep it under its sway and military presence. The Taliban were never a target in the war against terrorism.' They were just a scapegoat – rather, a horde of medieval warrior scapegoats who simply did not fulfil their contract: to insert Afghanistan into Pipelineistan.[81]

Zalmay Khalilzad, an influential voice in the US establishment, was working on risk analysis for Unocal at the Cambridge Energy Research Associates at the time. In a signed article in the *Washington Post*, he sought to allay fears about the Taliban. 'The Taliban does not practice the anti-US style of fundamentalism practiced by Iran – it is closer to the Saudi model. The group upholds a mix of traditional Pashtun values and an orthodox interpretation of Islam,'[82] the article sought to reassure the prospective TAPI countries and the investors. It not only gave a clean bill of health to Taliban but also portrayed the Saudi version of Islam as acceptable. The anti-US fundamentalism, on the other hand, was ascribed to Iran. Khalilzad was obviously seeking to put Iran out of the pipeline plans. The IPI was out, TAPI was in.

The TAPI plan was suddenly abandoned by the Unocal in 1998, when the bombing of two US embassies in Africa was linked to Osama bin Laden in Afghanistan.[83] The next year, the US president Bill Clinton banned US investments in Taliban-controlled areas in Afghanistan. None of these calamities was to dampen the Turkmen president Saparmurat Niyazov's hopes to build the pipeline. In May 1999, Turkmen and Taliban officials sought to revive the project amid Niyazov's continued efforts to mediate between the warring Afghan factions. Niyazov even asked the United Nations to consider supporting the Afghan pipeline project as a way of bringing stability to the country.

'If they have gone, others will come in. We are already being approached and talks are going on. Several companies want to come in place of Unocal,' the Turkmen oil minister Arazov Redjebai was confident.[84] The trans-Afghan pipeline project was also discussed at the May 2000 meeting between Niyazov and the Pakistani ruler General Parvez Musharraf.[85] And then came the infamous 9/11. Just a couple of days after the terrorist attacks on the US, the Unocal reiterated its complete dissociation from the project: 'We withdrew from that project in 1998, and do not have – nor plan to have – any projects in that country. We do not support the Taliban in any way.'[86]

The US war on Afghanistan revived the expectations regarding the pipeline. In view of the US determination to block the route through Iran, the Afghanistan–Pakistan corridor was the only alternative to unleash Central Asian energy potential. In May 2002, President Pervez Musharraf, Turkmenistan president Saparmurat Niyazov and Afghan interim prime minister Hamid Karzai signed an agreement in Islamabad to launch a feasibility study and look for ways of financing the pipeline. The summit talks also covered plans to lay regional gas and oil pipelines and build rail and road links to boost economic and trade cooperation among the three countries. Speaking on the occasion, Musharraf said Pakistan was developing the Gwadar port in south-western Balochistan Province to serve as a gateway for gas and oil exports to Japan and other countries.[87] It is interesting to note that Musharraf was keen to extend the IPI to Gwadar as well.

The post-war conferences on Afghanistan kept the focus on the TAPI pipeline. In 2006, at a donor meeting in New Delhi, countries promised to accelerate the planning of the pipeline and to help Afghanistan become an energy bridge. In 2008, at a meeting in Paris, Afghanistan's National Development Strategy (2009–2013) was presented to donors. The strategy mentioned ongoing planning for the TAPI gas pipeline and Afghanistan's central role as a land bridge connecting energy-rich Central Asia to energy-deficient South Asia. The Afghans strengthened their case by informing the meeting that more than a thousand industrial units were planned near the pipeline route and would need gas for their operation. They said 300 industrial units near the pipeline route had already been established, and the project's early implementation was essential to meet their requirements. In 2009, the Afghan government referred to the proposed pipeline again in documents relating to the First Afghan Hydrocarbon Bidding Round. The invitation to foreign companies to bid for exploration in the north of the country stated that 'the TAPI project . . . could be one of the export routes'.[88]

Afghanistan cherished its own ambitions to emerge as a gas-rich country and not just a transit one. Soviet estimates of the 1970s placed 'Afghanistan's "explored" (proved plus probable) gas reserves at about 5 trillion cubic feet (0.1415 tcm). The Hodja-Gugerdag's initial reserves were placed at slightly more than 2 tcf (0.0566 tcm)'.[89] The US Energy Information Administration (EIA) confirmed the Soviet figures much later. In 2008, it stated that Afghanistan's natural gas reserves were 'substantial'. As northern Afghanistan is a 'southward extension of Central Asia's highly prolific, natural gas-prone Amu Darya Basin', Afghanistan 'has proven, probable and possible natural gas reserves of about 5 trillion cubic feet (0.1415 tcm)'. Surprisingly, there was no discrepancy in the estimates given by the Soviets and the Americans.[90]

India joins the TAPI

It was nearly a decade after the TAPI pipeline was proposed that India got a green light to join it. In 2006, at the first meeting of the India–Turkmenistan intergovernmental commission on trade, economic, scientific and technological cooperation, the Turkmens offered to back India's entry into the project.[91] India was in the process of talks with the US on the civil nuclear energy and under severe pressure against joining the IPI. The TAPI project was an obvious way out of the situation.

In April 2008, the petroleum ministers of Turkmenistan, Afghanistan, Pakistan and India signed an agreement in Islamabad for the $7.6 billion TAPI.[92] The implementation would begin in 2010, and first supplies should start flowing through the 1,680-km pipeline in 2015. It was to carry 3.2 bcf (0.09 bcm) of gas a day. It would begin in Dauletabad and pass through Herat, Kandhar and Multan, ending at Fazilka on Indo-Pak border.[93] Speaking on the occasion, Deora said that both the TAPI and the IPI were equally important and India would want to settle and activate both because the demand in India was high and the prices were shooing up. Coming back home, he added that India and Pakistan had almost worked out a general agreement on the transit fee for the IPI.[94]

In September 2009, the foreign minister of India, S. M. Krishna, visited President Berdimuhamedov of Turkmenistan for discussions that included terms of the TAPI pipeline project. Berdimuhamedov paid a three-day visit to India in late May 2010. Five agreements on trade, commerce, culture and science were signed during the visit. The prominent event of the visit was the meeting of Deora with Berdimuhamedov and the visiting delegation,[95] suggesting an onward movement of the TAPI project.

In December 2010, a TAPI Summit was held in Ashgabat, where the four countries involved signed a gas pipeline framework agreement and intergovernmental agreement. The Indian Ministry of External Affairs, in its website, projected the event as promoting the bilateral ties to a strategic partnership between India and Turkmenistan.[96]

On 23 May 2012, the four countries signed the historic gas sales purchase agreement (GSPA) in Avaza, Turkmenistan. B. C. Tripathi, chairman of GAIL, and Sahatmurad Mamedove, the head of the Turkmengaz, signed the deal. The TAPI pipeline would have a capacity to carry 90 cm a day of gas for a 30-year period and was likely to become operational by 2018. India and Pakistan would get 38 cm each, while the remaining 14 cm would be supplied to Afghanistan. The contract price of the gas was linked to a formula which contained indices based on fuel basket and other indices which are not as volatile as crude oil. The formula was similar to the ones used in international contracts. A week earlier to the meeting, the Indian Union Cabinet had approved the signing of the agreement and the payment of 50 cents per million metric British thermal unit (mmBtu) as the transit fee to Pakistan and Afghanistan.[97]

After the change of government in India, the newly appointed minister of state for petroleum and natural gas Dharmendra Pradhan participated in the 19th Steering Committee meeting of the TAPI project in Ashgabat in November 2014. Since Turkmenistan does not provide for giving foreign firms an equity stake in the upstream gas field, the Western energy giants were not interested to take the risk. The American companies Chevron and Exxon Mobil and France's Total had withdrawn from the project after having expressed interest. India asked Turkmenistan to ease the rule to facilitate the construction process. The TAPI company has been set up to build, own and operate the pipeline in which Turkmengaz, Afghan Gas Enterprise, Inter State Gas System and GAIL own equal shares.

An extremely smart solution has since been found to move forward without insisting on amendments to the Turkmen law. Two consortiums are believed to be established. One will be a joint venture between the Turkmen government and Total for upstream operations, in which Total would serve as consortium leader with Russia's Rostec and the CNPC for the pipeline construction. While Turkmenistan would technically retain legal ownership of the land, Total probably has been given a sufficient profit share to warrant its assumption of the risk. Such a solution could come in the form of a modified technical services contract that would give Total the first right of refusal over gas extracted. The participation of Rostec and CNPC as consortium partners would be a savvy move to mollify Moscow and Beijing.[98] In April 2015, the Indian foreign minister Sushma Swaraj visited Turkmenistan to co-chair the fifth intergovernmental joint commission, where the TAPI project was also discussed.

The TAPI project was finally pulled out of the negotiating table onto physical terrain with the groundbreaking ceremony in Mary in the southeastern part of Turkmenistan on 13 December 2015. The ceremony was attended by Afghan president Ashraf Ghani, along with Turkmenistan president Berdimuhamedov, Pakistan prime minister Nawaz Sharif and Indian vice-president Muhammad Hamid Ansari. Its strategic significance is seen to be huge. The project is seen to be a game changer in regional geopolitics and economic integration.[99] It would accelerate India's 'Connect Central Asia' policy. It would help promote Pakistan's 'China–Pakistan Economic Corridor' (CPEC) and the Chinese 'One Belt, One Road' project. The transit revenue to Afghanistan could provide for the country's economic reconstruction. The great question mark hanging on the project's progress is security. The Afghan government has decided to raise a security force to guard the line, and Pakistan is understood to have agreed to use its influence with the Taliban to keep the line trouble-free.

A year earlier, the Russian president Putin had discussed the possibility of building a pipeline along the route of the TAPI, which would run from Russia's southern border to India. The route through the Himalayas could also be possible, according to the Russian ambassador in India, Alexander Kadakin. The line would be the 'biggest ever energy project in history', he says.[100] The offer is at the very initial stage and could mature in time. It is not clear as to if the suggestion is to complement the TAPI project or substitute it.

Myanmar-Bangladesh-India Pipeline

It is noteworthy that when the Unocal was preparing to launch the TAP,[101] it was also planning to launch a pipeline from Bangladesh to India. Together,

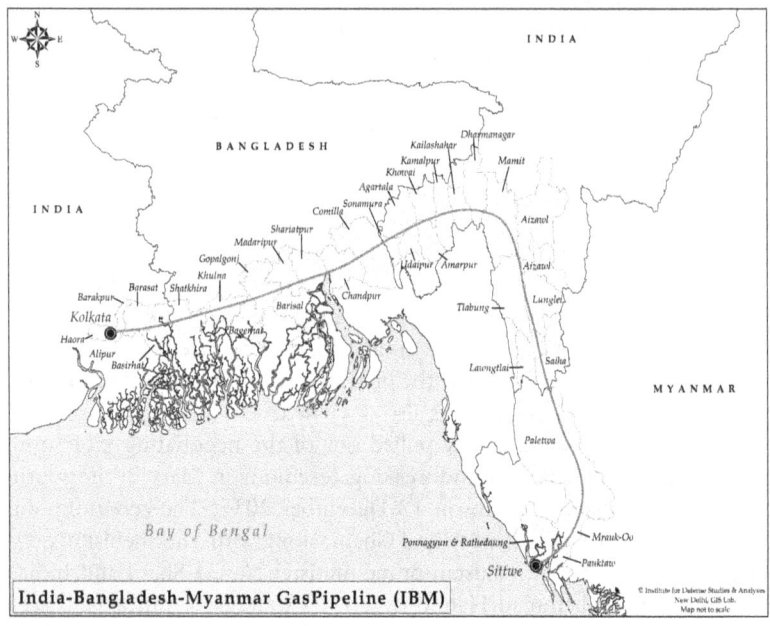

Figure 6.3 India–Bangladesh–Myanmar pipeline.

Source: Institute for Defence Studies and Analyses.

Disclaimer: Map not to scale.

the two would have constituted the South Asia Integrated Gas (SAIG) project.[102] The Unocal was also in the process of setting up a $2 billion integrated pipeline connecting Myanmar to the gas fields of Bangladesh on to Haldia in India. In reverse energy flow Bangladesh would be fed with natural gas from the fields of Tripura. As if the two projects were interlinked (they should not have been), the TAP project seemed to have taken Myanmar–Bangladesh–India (MBI) with it, causing Unocal to withdraw from both.

Around that time, the Indian decision-making institutions seemed to be of the view that there were four important economic compulsions on Bangladesh that would lead to a decision in favour of export of gas to India: (1) absence of any other viable market; (2) contractual commitments with the International Oil Companies under which Bangladesh government had to make payments in foreign currencies; (3) balance of payment problems faced by the Bangladeshi government; and (4) decline in garment exports, which was leading the export sector.[103]

The MBI, therefore, had support of a favourable estimate within the Indian External Affairs Ministry. In addition, it answered the Indian energy security policy of diversification from an overwhelming dependence on West Asia. India agreed to sign a trilateral deal with Bangladesh and Myanmar in early 2005 to build the 290-km pipeline. Its expected cost of $1 billion was to be mostly borne by India. If the plan had gone ahead, Bangladesh had expected to get about $350 mm in investment and to earn $100 mm in annual transmission fees. Bangladesh also expected to get another $100 mm as one-off right-of-way charges from the project and $25 mm each year for sharing management.[104] It needs to be noted that it was the first-ever gas pipeline deal that India entered into. The TAPI came later, and the IPI was never formally signed and sealed.

The Bangladesh–India bilateral tensions were the primary hindrances on the ground level. Dhaka was aware of India's growing need for the gas and raised its demand accordingly. It set forth a number of conditions for letting its territory be used for transit: establishing trade routes for commodities from Bangladesh to Nepal and Bhutan through Indian territory; allowing transmission of hydroelectricity from Nepal and Bhutan to Bangladesh through Indian territory; and pursuing measures to reduce Bangladesh's trade imbalance with India.[105] India was reluctant to take extraneous factors on-board and insisted on negotiating strictly pipeline-related issues. The asylum to the militants from the Indian North-East did not help in generating trust between the two.

The Bangladeshi position had a few domestic components of its own. The National Energy Policy released in 2004 identified three policy aims as exploitation of indigenous energy sources, diversification of energy type and tapping into the lowest cost fuels available.[106] Thus, if India prioritised diversification of supply, so did Bangladesh wish to secure diversified export markets. India was negotiating two pipeline projects, IPI and TAPI, at that time. Bangladeshi fears that India might ignore or underplay the Bangladesh connection were justified, in the circumstances.

Finally, there were domestic political considerations on both sides. The anti-India voices in Bangladesh were dead-set against such a long-term commitment with it.[107] The Indian political establishment did want to enter into a deal that could have been a manna to the anti-India BNP government ruling the country.

In view of the stubborn logjam, India turned to Myanmar. Bangladesh would be completely left out of the pipeline map; instead, the route would be through India's North-east. Three possible pipeline routes for Myanmar gas to reach the Indian market were considered. The first was to lay the pipeline along the Kaldan River of Myanmar, which has tributaries in India's Mizoram area in the North-east. This pipeline could then be linked up with India's national oil and gas grid passing through Assam and Meghalaya states. The other route could be an offshore connection from

Sittwe (Akayab) port in Myanmar, near the A-1 PSC exploration block, to some point in West Bengal state. From there, it could then be linked to the national network or the designated consumption points.

The third route was through Bangladesh, linking the gas field in Myanmar with the pipeline in Bangladesh, laid to carry gas from Bangladesh to India.[108] Skirting around Bangladesh would have inflated the cost to $3 billion due to longer distance, expected to cover a distance of 1,400 km. Logistically, gas from Myanmar could not bypass Bangladesh as bringing a pipeline down from Yangon and Yeravati delta and bringing it up again at Bay of Bengal, though may be technically feasible, did not appear to be economically a viable option.[109]

By then, Myanmar was already in negotiations with China and Thailand for a gas pipeline. In December 2005, it signed a deal with China for a project that consisted of dual oil and gas pipelines originating in Kyaukryu on the west coast of Myanmar entering China at Yunnan's border city of Ruili.[110] Myanmar assured India that gas could still be made available. India hired the Brussels-based consulting firm Suz Tractebel to conduct a feasibility study for overland pipeline routes to North-east India, circumventing Bangladesh territory.[111] In March 2007, the Myanmar government announced that it was not prepared to export gas by pipeline to India, or even as LNG, but preferred instead the Chinese offer to build a 900-km pipeline to the Chinese border. The new pipeline would have a total length of 2,380 km extending from Sittwe to the Chinese city of Chongqing.[112]

The MBI chapter is still on the crutches limping along. A few years back, Bangladesh was lured by the prospect of entering the TAPI, once India became a stakeholder in it. The energy experts inside the South Asia Association for Regional Cooperation (SAARC) have been discussing the possibility. The pro-India Awami League party that came into power in Bangladesh in 2010 has approved the pipeline project. India too has stepped up its diplomatic efforts and, at the risk of upsetting the sitting prime minister Sheikh Hasina, has been courting the opposition Bangladesh Nationalist Party (BNP) leader Begum Khaleda Zia.[113]

SAGE energy corridor

The Middle East to India Deep Water Pipeline (MEIDP) is an initiative by the South Asia Gas Enterprises Private Limited (SAGE), which is promoted by the Delhi-based Sidhomal Group. The pipeline would be 1,300 km in length and would cross the Arabian Sea at the maximum depth of 3,400 metres. It would bring 8 tcm of gas to India over the next 20 years. The gas would be sourced from Qatar, Iran, Iraq and Turkmenistan with an option of gas swap between them. On route to India, gas would also be supplied to Oman and the UAE. The supplier states would have a right to use the pipelines on payment of tariff. Over the next 10 years, two more lines would be built.[114]

Two route options were being considered for laying the line: one, a single uniformly sized pipeline from Ras al-Jifan in Oman to Gujarat in India; and two, the same route via a midline offshore compression station on the Qualhat Seamount.[115] The depth of the water is reduced to approximately 300 metres at this point.

At the Second World Energy Policy Summit held in New Delhi on 27 December 2012, Ian Nash, one of the key team members of the SAGE made a presentation titled 'The Middle East to India Gas Pipeline: Challenges and Opportunities'. He listed four arguments in favour of the MEIDP: one, over 2,000 tcf (56.6 tcm) of natural gas reserves are held by countries with which India has a traditional trading relationship, including Qatar, Iran and Turkmenistan; two, onshore pipelines such as the IPI and TAPI have significant security and supply issues; three, Qatar is looking for new export markets with the advent of shale gas explosion in US; and four, Iraq is building its gas development and looking for export solutions.[116]

Suboth Kumar Jain, the director of the Sage and the Sidhomal Group, put the estimated cost at $5 billion, the time frame at three to four years for the completion of the project and the supply of 31 cm a day of gas to India.[117] He said that SAGE had made presentations to the Kelkar Committee,[118] the Ministry of Petroleum and Natural Gas, the planning commission and that they all thought it was a doable project.[119] The technical and commercial feasibility reports for the pipeline were completed and INTECSEA[120] of UK/US confirmed that the project is technically feasible.[121]

At the India end, there was a quiet confidence in the success of the venture. In August 2009, GAIL signed an agreement with the SAGE. A senior official from GAIL, was quoted saying,

> We expect that the pipeline will draw good investment response from players in India and West Asia. GAIL has procured intensive studies conducted by leading oil and gas industry deepwater pipeline specialists, Heerema Marine Contractors and INTECSEA, who have said that the Mideast pipeline project is technically feasible. This was an important consideration in signing the deal with SAGE.[122]

There was a media report suggesting that a strong lobby of LNG suppliers like Chevron, Exxon Mobile, Shell and Conoco were not in favour of the piped gas even though the landed price of that gas would be $10 mmBtu as compared to the price of $13 to 14 mmBtu that India was paying for the LNG.[123] There were informed voices expressing doubts on account of the deal being between national oil companies, in which the bureaucrats with no entrepreneurial background were negotiating the contractual terms. The BTC and Shah Deniz, on the other hand, were carried out by the private oil companies keen to monetise their oil and gas reserves. The state-operated pipeline projects have remained pipe dreams precisely for this reason; so did the argument go.[124]

At the Iran end, the response was equally positive. It coincided with the cancellation of its $500 million loan to Pakistan to build part of its pipeline. The Deputy Oil Minister Ali Majedi said the cash-strapped Iran was not obliged to finance Pakistan side of the pipeline. Ali Amirani, the director of marketing at the National Iranian Gas Exports Company (NIGEC) informed of the negotiations held with three Indian companies and agreements reached. Bijan Namdar Zanganeh, the oil minister, identified India as one of the destinations, which could be linked with cost-effective pipelines.[125] Thus, the entire Iranian hierarchy of gas-related officials expressed its willingness to go ahead with the project.

A decisive development on the ground took place on 28 February 2013, when the Indian minister of external affairs Salman Khurshid met with the Iranian foreign minister Jawad Zarif and the Omani minister responsible for foreign affairs Yusuf bin Alawi bin Abdullah. There was a general agreement on the deep water pipeline. Iran would supply gas from the South Pars field, according to reports from the Iranian Press TV. In addition, Zarif informed that Iran was negotiating separately with Turkmenistan for an overland pipeline to carry its gas to an Iranian terminal and thence to the markets like in India.[126] If the things work out, then Iran would not just be a supplier itself, but a hub to collect gas from its neighbourhood and transport it to markets. Zanganeh had already identified India among the countries, where it would be cost-effective to supply gas by pipelines.[127]

Oman pursued the matter with Iran and India bilaterally as well. In March 2014, it signed a 'Heads of Agreement' with Iran during President Hasan Rouhani's visit to Muscat.[128] In his talks with the Indian foreign minister Sushma Swaraj, Yusuf bin Allawi showed interest in reviving the long-stalled Oman–India undersea pipeline.[129] India had signed an in-principle agreement with Oman way back in 1994. In order to avoid the territorial waters and the EEZ of Pakistan, a deep-sea route was proposed, parts of which lay at the water depths exceeding 3,000 metres. At that time it was felt that there was no appropriate technology to lay pipelines at such depths and with deep and sudden changes in sea bed terrain. There was also a question mark as to whether the proven gas reserves in Oman would be able to support all the projects being planned. As a consequence, this pipeline proposal was not pursued further.[130] Oman has since been import-ing gas from Qatar.

There was an unexpected deal breaker, just when things were in a smooth sail. On 15 March 2015, the United Nations Commission on the Limits of the Continental Shelf formally accepted Pakistan's case for the extension of its continental shelf from 200 nautical miles to the maximum 350 nautical miles. The total maritime area gained by Pakistan was more than 50,000 sq. km, as a result.[131] What then are the implications of this development on the prospects of the SAGE pipeline?

Pakistan now has an exclusive right over the resources in and under its EEZ, that is marine resources in water and mineral and metal resources under the water. The SAGE line will be able to traverse the area only if Pakistan permits.[132] The pipeline has tabled two routes, one of which plans to build a midline offshore compressor station on the Qualhat Seamount, which would now fall under the Pakistan control. The Muscat Agree-ment on the delimitation of the Maritime Boundary between Pakistan and Oman[133] deals specifically with the geological petroleum structure and provides for a dispute resolution mechanism. Oman, an interested party in the SAGE project, may take recourse to this mechanism. Even if the SAGE decides to use the route that would go straight from Ras al-Jifan to India, the pipeline would still need to cross the Pakistan EEZ. In short, either permission from Pakistan or an inclusion of Pakistan is the only way forward.

Recapitulation and leads

India is surrounded by gas-rich countries: Myanmar and Bangladesh to the east, the Gulf to the west and Central Asia to the north. The development of natural gas industry in India started in 1960s, when gas fields were dis-covered in Assam and Gujarat. More discoveries have followed, and there is an expectation that the Indian continental shelf deep water may contain recoverable hydrocarbons. As the demand is continuously going up, gas import is the only available option till then. At the apex of the energy structure is the Ministry of Petroleum and Natural Gas (MoPNG). The Gas Authority of India Limited (GAIL) was incorporated as a PSU in the mid-1980s. Petronet LNG Limited, a joint venture company, and the Reliance Industries (RIL), a privately owned company, are other important actors in the gas sector.

The IPI gas pipeline project has been under consideration for more than two decades. The talks have repeatedly broken down over transit fee issue between India and Pakistan and over pricing issue between India and Iran. The US sanctions on Iran, the US pressure on India against joining the project and the Indian considerations of security of the line have left the project in hibernation.

Originally conceived as TAP, the line was planned to deliver gas from Turkmenistan to Pakistan via Afghanistan. It is now named TAPI after India joined the project in 2006. India was in the process of talks with the US on the civil nuclear energy and under severe pressure against joining the IPI. The TAPI project was an obvious way out of the situation. The TAPI company has been set up to build, own and operate the pipeline in which Turkmengaz, Afghan Gas Enterprise, Pakistan's Inter State Gas System and GAIL own equal shares. The groundbreaking ceremony in December 2015 has finally set the pipeline on the move.

India has sought to source the gas from its east as well. A Unocal initiative originally, India, Bangladesh and Myanmar signed a tripartite agreement in early 2005. By the end of the year, Bangladesh leg of the tripod had collapsed. India sought to chart out a route over the Indian North-east to access Myanmar gas. In 2007, Myanmar withdrew from its earlier assurances, preferring to sell to China instead. The project might still be revived in any of the various routes considered earlier.

The MEIDP, an initiative by the SAGE, seeks to succeed where the previous ones failed. The pipeline would be 1,300 km in length and would cross the Arabian Sea at the maximum depth of 3,400 metres. It would bring 8 tcm of gas to India over the next 20 years. Pakistan has since acquired an extension of its continental shelf from 200 nautical miles to the maximum 350 nautical miles. The MEIDP would have to traverse the Pakistan EEZ, as a result. In this situation, either permission from Pakistan or an inclusion of Pakistan in the project is the only way forward.

Notes

1 *India Hydrocarbon Vision*, 2015. http://www.petroleum.nic.in/vision. doc. Accessed on 23 February 2015.
2 *Report of the Group on India Hydrocarbon Vision 2025* (RGIHV, 2025), p. 81.
3 B. K. Chaturvedi, 'Domestic Resourcing of Energy, Gulf and the Future of Global Energy', in Dr. S. N. Malakar, ed., *India's Energy Security and the Gulf* (Academic Excellence, Delhi, 2006), p. 16.
4 http://petroleum.nic.in/docs/abtng.pdf. Accessed on 3 July 2014.
5 Hippu Salk Kristle Nathan, Sanket Sudhir Kulkarni and Dilip R. Ahuja, 'Pipeline Politics – A Study of India's Proposed Cross Border Gas Projects', *Energy Policy*, Vol. 62, 2013, p. 145.
6 Quoted in 26 June 2014. http://www.eia.gov/beta/international/analy sis.cfm?iso=IND. Accessed on 23 May 2015. The British Petroleum puts the figure at 1.4 tcm. See Chapter 1.

7 Najeeb Jung, 'Natural Gas in India', in Wybrew-Bond Ian and Jonathan Stern, eds., *Natural Gas in Asia: The Challenges of Growth in China, India, Japan and Korea* (Oxford University Press, New York, 2002), p. 96.

8 Chaturvedi, n. 3, p. 14.

9 The tonne of oil equivalent (toe) is a unit of energy: The amount of energy released by burning one tonne of crude oil. Different crude oils have different calorific values, and, therefore, the exact value and definition of the toe is different.

10 http://www.iea.org/statistics/statisticssearch/report/?country=INDIA& product=naturalgas&year=2012. Accessed on 13 December 2013.

11 http://www.bp.com/en/global/corporate/about-bp/energy-econom ics/statistical-review-of-world-energy/review-by-energy-type/natural-gas/natural-gas-production.html. Accessed on 13 December 2013.

12 http://www.bp.com/en/global/corporate/energy-economics/energy-outlook-2035/country-and-regional-insights/india-insights.html. Accessed on 6 January 2016.

13 http://www.bp.com/en/global/corporate/about-bp/energy-econom ics/statistical-review-of-world-energy/country-and-regional-insights/ india-insights.html 2014. Accessed on 13 December 2013.

14 http://petroleum.nic.in/docs/abtng.pdf. Accessed on 13 December 2013.

15 *India Hydrocarbon Vision*, n. 1.

16 *Report of the Group on India Hydrocarbon Vision*, n. 2, pp. 62–63.

17 Jonathan Stern and Wybrew-Bond Ian, 'Conclusions', in Wybrew-Bond Ian and Jonathan Stern, eds., n. 7, p. 305.

18 'Natural Gas Vehicle Statistics', *International Association for Natural Gas Vehicles*. http://www.iangv.org/tools-resources/statistics.html. Accessed on 13 December 2013. Quoted in Charles K. Ebinger, *Energy and Security in South Asia: Cooperation or Conflict?* (Brookings Institute Press, Foundation Books, Delhi, 2013), p. 43.

19 *India Hydrocarbon Vision 2025*, n. 1.

20 *RGIHV*, 2025, n. 2, p. 22.

21 'Strengths, Weakness, Opportunities and Threats and Challenges Analysis', *Strategic Plan for 2011–2017*, pp. 12–14. http://petroleum.nic.in/docs/ reports/stratreport.pdf. Accessed on 23 January 2013.

22 Rahul Roy-Chaudhury, 'An Energy Security Policy for India: The Case of Oil and Natural Gas', *Strategic Analysis*, February 1998, pp. 1671–1672.

23 *Integrated Energy Policy: Report of the Expert Committee* (Planning Commission, Government of India, 2006). http://planningcommission.nic.in/ reports/genrep/rep_intengy.pdf. Accessed on 3 January 2007.

24 http://petroleum.nic.in/docs/abtng.pdf. Accessed on 23 January 2013.

25 The government has designated seven public-sector undertakings as Maharatnas (Great Pearls, in literal translation) on three criteria: net worth, net profit and net turnover. Below the maharatnas are navratnas and then miniratnas.

26 http://www.gailonline.com/final_site/index.html. Accessed on 23 January 2013.

27 http://www.pngrb.gov.in/newsite/about-us.html. Accessed on 25 January 2014.

28 http://www.iocl.com/AboutUs/Profile.aspx. Accessed on 27 January 2014.

29 http://www.oil-india.com/Pipelines.aspx. Accessed on 27 January 2014.

30 'IOC Seeks Nod to Join Deep Sea Gas Pipeline Project of SAGE', *The Hindu*, 24 June 2012.

31 http://www.petronetlng.com/. Accessed on 27 January 2014.

32 http://www.ril.com. Accessed on 23 March 2014.

Paranjoy Guha Thakurta with Subir Ghosh and Jyotirmoy Chaudhari, *Gas Wars: Crony Capitalism and the Ambanis* (Paranjoy Guha Thakurta Publishers, New Delhi, 2014) gives a great insightful account of the controversy.

33 18 March 2013. http://www.eia.gov/countries/cab.cfm?fips=IN. Accessed on 24 March 2014.

34 'India: Many Discoveries Later, Gas Is Still a Pipedream', *Gas and Oil*, 11 July 2009. http://www.gasandoil.com/news/2009/09/nts93531. Accessed on 13 March 2014.

35 Promit Mukherjee, 'India Likely to Miss 12th Plan Target on Gas Pipelines', *Livemint*, 6 June 2014.

36 http://www.infraline.com/ong/reforms/vision2025-ii.pdf, p. 20.

37 Leena Srivastava and Megha Shukla, 'The Present and Future Prospects of Energy in India', in *Asian Energy Markets: Dynamics and Trends* (Emirates Center for Strategic Studies and Research, Abu Dhabi, 2004), p. 335.

38 'India More Than Doubling Natural Gas Pipeline Capacity', *PennEnergy*, 3 April 2012. http://www.pennenergy.com/index/petroleum/display/7205185667/articles/pennenergy/petroleum/refining/2012/april/india-more_than_doubling.html. Also, *Economic Times*, 2 April 2012.

39 *Free Press Journal*, 25 August 2014.

40 Ahmad Rashid, *Taliban: Militant Islam, Oil and Fundamentalism in Central Asia* (Yale University Press, New Haven, 2000), pp. 168–169.

41 The delivery of Siberian energy to South Korea presents a similar conundrum: whether to ship the oil and gas via North Korea – the most practical route but one that raises obvious national security considerations for Seoul – or through costly underwater conduits in the Yellow Sea. Michael T. Klare, 'Geopolitics Reborn: The Global Struggle for Oil and Gas Pipelines', *Current History*, Vol. 103, no. 677, 2004, p. 432.

42 *Observer of Business and Politics* (New Delhi), 13 February 1997.

43 C. Raja Mohan, 'Positive Outlook for Indo-Iranian Pipeline', *The Hindu* (New Delhi), 24 November 2000.

44 Hooman Peimani, 'An Economic and Political Pipeline', *Asia Times*, 20 November 2002.

45 B. Muralidhar Reddy, 'Pakistan Ready to Give Guarantees for Gas Pipelines', *The Hindu*, 8 June 2004.

46 http://meaindia.nic.in. Accessed on 22 April 2005.

47 Neil Buckley, Francesco Guerrera, Ray Marcelo and Kevin Morrison, 'India Looks to Russia and Iran for Energy', *Financial Times* (London), 8 January 2005.

48 M. K. Bhadrakumar, 'India Finds a $40bn Friend in Iran', *Asia Times*, 11 January 2005.

49 'Indo-Iran Gas Pipeline Can Be Pulled to Southern China via Burma', *Indian Express*, 15 February 2005.

50 'There Shall Be Freedom of Transit through the Territory of Each Contracting Party, via the Routes Most Convenient for International Transit, for Traffic in Transit to or from the Territory of Other Contracting Parties.

No Distinction Shall Be Made Which Is Based on the Flag of Vessels, the Place of Origin, Departure, Entry, Exit or Destination, or on Any Circumstances Relating to the Ownership of Goods, of Vessels or of Other Means of Transport', *World Trade Organisation*. http://www.wto.org/english/res_e/booksp_e/gatt_ai_e/art5_e.pdf. Accessed on 30 March 2014.

51 'India Wants WTO Shield for IPI Gas Pipeline', *Gas and Oil*, 24 August 2005. http://www.gasandoil.com/news/2005/09/nts53770. Accessed on 30 March 2014.

52 Bhadrakumar, n. 48.

53 The Obama administration has continued the policy followed during the Bush presidency. Rice's successor, Secretary of State Hillary Clinton, justified a $1 billion dollar assistance, with energy infrastructure projects in Pakistan before a House Appropriations Subcommittee on Foreign Operations by stating that the requested funds were aimed to assist with a gas pipeline project from Turkmenistan to Pakistan via Afghanistan. The alternative, she feared, was that Pakistan would choose to buy gas from a pipeline being built by Iran with financial support from the Chinese Export–Import Bank.
   Salim H. Ali, 'Pipeline Diplomacy with Iran, Pakistan', 7 March 2012. http://www.huffingtonpost.com/saleem-h-ali/pipeline-diplomacy-with-iran-pakistan_b_1325664.html. Accessed on 12 March 2012.

54 B. Muralidhar Reddy, 'US, India Can Explore Alternatives', *The Hindu*, 19 March 2005.

55 B. Muralidhar Reddy, 'Pakistan Studying Four Options to Acquire Natural Gas', *The Hindu*, 4 April 2005.

56 *The Hindu*, 7 August 2007.

57 His appointment replacing Mani Shankar Aiyar was seen as an indication of India's changing stand on the IPI and developing closeness to the US.

58 *The Hindu*, 1 November 2007.

59 He said that at a meeting in the Institute for Defence Studies and Analyses, New Delhi, on 15 February 2007. The author was present at the meeting.

60 *The Hindu*, 30 April 2008.

61 Peter Kiernan, 'Pipeline Politics: Iran Looks to Send Natural Gas East', *World Politics Review*, 28 May 2008.

62 *The Hindu*, 28 April 2008.

63 *The Hindu*, 21 April 2008.

64 *The Hindu*, 29 April 2008. The 1,300 kilometre-long Karakoram Highway is the only overland link for China to its strategic prize in Pakistani port of Gwadar, and a threat to India.

65 Even if the Indian Industry and consumers are willing to pay a higher price, 'the Indian defence, foreign and security establishments strongly believe that the time is not ripe for international transit pipelines', according to Narendra Taneja, an energy expert. Quoted in Ashwani Talwar, 'India: Pipelines to Nowhere?', *Gas and Oil*, 24 May 2008. http://www.gasandoil.com/news/2008/06/ntm82553. Accessed on 31 March 2014.

66 Gurmeet Kanwal, 'IPI Pipeline Is a Good Option, but Security Is a Nightmare', *Gas and Oil*, 6 July 2008. http://www.gasandoil.com/news/2008/08/ntm83256. Accessed on 24 April 2014.

67 Swaminathan S. Anklesaria Aiyar, 'Pipeline Lessons for India from Ukraine', *Gas and Oil*, 7 January 2006. http://www.gasandoil.com/news/2006/01/nts60461. Accessed on 27 April 2014.

68 Rahul Tongia, 'Economics of IPI Pipeline', *Gas and Oil*, 3 February 2007. http://www.gasandoil.com/news/2007/02/nts70996. Accessed on 2 December 2014.

69 Harsh Pant, 'The Iran Factor in India-Gulf Ties: An Indian Perspective', in Ranjit Gupta, Abu Backer Bagader, Talmiz Ahmad and N. Janardhan, eds., *India and the Gulf: What Next?* (Gulf Research Centre, Cambridge, 2013), p. 154.

70 Amitav Ranjan, 'India to Step up Pressure for Iran Gas Line', *Indian Express*, 1 November 2008.

71 'India Can Bypass Pakistan for Gas Pipeline Project with Iran: ASSOCHAM Study', 30 August 2015. http://www.assocham.org/newsdetail.php?id=5176. Accessed on 30 September 2015.

72 Shebonti Ray Dadwal, 'Need to Revive India-Iran Ties', 30 September 2015. http://idsa.in/idsacomments/NeedtoreviveIranIndiaenergyties_srdadwal_300914.html. Accessed on 1 October 2015.

73 M. K. Bhadrakumar, 'India Stranded as Region Readies for Iran's Surge', *Asia Times*, 9 April 2015.

74 Frederic Grare, 'Meeting India's Energy Needs: What Role for Central Asia?', in Pierre Audinet, P. R. Shukla and Frederic Grare, eds., *India's Energy: Essays on Sustainable Development* (Manohar Publishers, New Delhi, 2000), pp. 247–248.
   A picturesque description of the TAPI route goes like this: Imagine a steel serpent entering western Afghanistan towards Herat, going south underground (to prevent terrorist bombing) parallel to the Herat-Kandahar road, then taking a detour via Quetta – home of Taliban supremo Mullah Omar – to Multan in Pakistan and finally reaching Fazilka, on the Indian border. Pepe Escobar, 'Playing Chess in Eurasia', *Opednews*, 25 December 2011. http:opednews/articles/ Playing-Chess-in-Eurasia-by-Pepe Escobar-111225-365.html. Accessed on 11 January 2013.

75 'Unocal Bharat Limited', *Energy Solutions for India* (New Delhi, March, 1998), p. 3.

76 A Tarock, 'The Politics of the Pipeline: Iran and the Afghanistan Conflict', *Third World Quarterly*, Vol. 20, no. 4, 1999, p. 817.

77 R. Hrair Dekmejian and Hovann H. Simonian, *Troubled Waters: The Geopolitics of the Caspian Region* (I. B. Tauris, London, 2001), p. 100.

78 Bruce Pannier, 'Turkmen, Iranian Presidents Moving Ahead with Rival Pipelines', *RFL/RE*, 28 April 2008.

79 John Foster, 'Afghanistan, the TAPI Pipeline, and Energy Geopolitics', *Journal of Energy Security*, 23 March 2010.

80 Pepe Escobar, 'Pipelineistan, Part 1: The Rules of the Game', *Asia Times*, 25 January 2002.

81 Pepe Escobar, 'Pipelineistan, Part 2: The Games Nations Play', *Asia Times*, 31 January 2002.

82 Zalmay Khalilzad, 'Afghanistan: Time to Reengage', *Washington Post*, 7 October 1996. Khalilzad later rose to become ambassador to Afghanistan, where he was a virtual ruler.

83 By that time, the Unocal had opened a school in Kandahar to train workers in pipe-fitting, welding and other skills required for the project. When Unocal announced it was suspending its involvement in the bidding in August 1998, the company denied it had trained Afghan workers for the project. Paul Watson, 'Afghanistan Aims to Revive Pipeline Plans', *Los Angeles Times*, 30 May 2002.

84 Sreedhar and Mahendra Ved, *Afghan Buzkashi: Power Games and Gamesmen* (Wordsmiths, Delhi, 2000), pp. 254–255.
85 Dekmejian and Simonian, n. 77, p. 101.
86 24 December 2001. http://www.unocal.com. Accessed on 30 September 2008.
87 *Gulf News*, 31 May 2002.
88 John Foster, 'Afghanistan, the TAPI Pipeline, and Energy Geopolitics', *Journal of Energy Security*, 23 March 2010.
89 'The Soviet Union to Retain Influence in Afghanistan', *Oil and Gas*, 2 May 1988.
90 John C. K. Daly, 'Analysis: Afghanistan's Untapped Energy', *UPI*, 24 October 2008.
91 Indian Ministry of External Affairs. Quoted in 'Turkmenistan Backs India's Bid to Join Gas Pipeline Project', *Dow Jones Newswire*, 5 October 2006.
92 The initial estimate of $2 billion had gone up considerably by the time India formally joined the plan. By 2015, the cost estimate had further appreciated to $10 billion, according to the Asian Development Bank. *Natural Gas Asia*, 11 February 2015. http://www.naturalgasasia.com/cost-of-tapi-gas-pipeline-could-rise-to-10-bn-14755. Accessed on 15 May 2015.
93 *The Hindu*, 25 April 2008.
94 *The Hindu*, 28 April 2008.
95 Aurobinda Mahapatra, 'The Discussion on the TAPI Gas Pipeline Project Would Continue', *Oriental View*, 1 June 2010. http://orientalreview.org/2010/06/01/the-discussions-on-the-tapi-gas-pipeline-project-would-continue/. Accessed on 11 January 2014.
96 *Ministry of External Affairs*, June 2012. http://www.mea.gov.in/Portal/ForeignRelation/Turkmenistan-June-2012.pdf. Accessed on 30 January 2013.
97 'TAPI Pipeline Gas Sale Agreement Signed', *The Hindu*, 23 May 2012.
98 Michael Tanchum, 'A Breakthrough on TAPI Pipeline?', *The Diplomat*, 20 March 2015. http://thediplomat.com/2015/03/a-breakthrough-on-the-tapi-pipeline. Accessed on 13 May 2015.
99 Gulshan Sachdeva, 'TAPI Gas Pipeline May Become Game Changer in South Asian Geopolitics', *Hindustan Times*, 20 December 2015.
100 *The Hindu*, 21 July 2014.
101 For more on this, see Chapter 5.
102 *The Economic Times*, 9 December 1997.
103 'Based on a Note prepared by the BSM Division of the MEA', 19 September 1999 in S. D. Muni and Girijesh Pant, *India's Search for Energy Security: Prospects for Cooperation with Extended Neighbourhood* (Rupa and Company, New Delhi, 2005), p. 88.
104 *Gas and Oil*, 2 August 2006. http://www.gasandoil.com/news/2006/08/nts63470. Accessed on 13 May 2007.
105 'Primary Jolt in Burma-India Gas Supply Project', *Mizzima News*, 19 January 2005. Quoted in Deepak V. Ganapathy, 'Hissing Dragon-Squirming Lion: China's Successful Strategic Encirclement of India', *South Asia Analysis Group*, Paper 682, 2003.
106 Varigonda Kesava Chandra, 'The Pipeline That Wasn't: Myanmar–Bangladesh–India Natural Gas Pipeline', *Journal of Energy Security*, 19

April 2012. http://www.ensec.org/index.php?option=com_content&vie w=article&id=348:india-bangladesh-and-the-myanmar-bangladesh-india-natural-gas-pipeline-how-not-to-achieve-energy-s&catid=123:content& Itemid=389. Accessed on 3 January 2014.

107 The politics in Bangladesh was preventing any kind of contract with India to sell its own gas and not allowing transit routes for gas from Myanmar. Bhamy V. Shenoy, 'Why India's Import Pipelines Have Remained Pipe Dreams', *Gas and Oil,* 7 January 2007. http://www.gasandoil.com/ news/2007/01/nts70531

108 Muni and Pant, n. 103, pp. 148–149.

109 Hippu Salk Kristle Nathan, Sanket Sudhir Kulkarni and Dilip R. Ahuja, 'Pipeline Politics- A Study of India's Proposed Cross Border Gas Projects', *Energy Policy,* Vol. 62, 2013, pp. 145–156.

110 Chandra, n. 106.

111 Syed Ali Mujtaba, 'Consultant to Study Indo-Burma Gas Pipeline Routes', *Mizzima News,* 8 February 2006. Quoted in Ganapathy, n. 105.

112 'China Plans Major Investment in Oil, Gas Pipeline to Myanmar', *Pipeline Asia,* 24 April 2007.

113 Daaman Thandi, 'Myanmar Pipedream: Myanmar-Bangladesh-India Pipeline', *Indian Review of Global Affairs,* 26 December 2013. http:// www.irgamag.com/analysis/terms-of-engagement/item/6622-the-myanmar-pipedream-myanmar-bangladesh-india-pipeline. Accessed on 3 May 2015.

114 http://www.sage-india.com/index.php?option=com_content&view=arti cle&id=48&Itemid=54. Accessed on 5 May 2015.

115 Ibid.

116 Ian Nash, 'The Middle East to India Gas Pipeline: Challenges and Opportunities' (2nd World Energy Policy Summit, 27 December 2012, New Delhi). http://www.academia.edu/2280732/Middle_East_to_India_ Deepwater_Pipeline_Challenges_and_Opportunities. Accessed on 14 May 2015.

117 Indrani Bagchi, 'India, Iran and Oman Go under Sea to Build Pipelines, Change Geopolitics', *Times of India,* 1 March 2014. See also 25 September 2013. http://www.projectsmonitor.com/daily-wire/will-oman-india-gas-pipeline-be-a-reality/. Accessed on 21 December 2013 and Monish Gulati, 'India-Oman Gas Pipeline: Not Just a Pipedream', *South Asia Monitor,* 20 March 2014. http://southasiamonitor.org/detail. php?type=sl&nid=7615. Accessed on13 May 2014.

118 The committee was mandated to work out a roadmap for fiscal consolidation under the chairmanship of Vijay Kelkar.

119 14 February 2014. http://www.dnaindia.com/money/report-now-sage-deepwater-pipeline-comes-into-focus-1961929. Accessed on 22 December 2014.

120 INTECSEA is an international company dealing with offshore field development and pipeline projects.

121 'Safe Deepwater Gas Pipeline from West Asia to India to Be a Reality', *Gas and Oil,* 6 June 2009. http://www.gasandoil.com/news/2009/08/ ntm93111

122 'A Qatar-to-India Pipeline?', *Gas and Oil,* 3 September 2009.

123 Ibid, 1 MMBTU ≈ 27.096 cm.

124 Shenoy, n. 107.
125 Atul Aneja, 'Iran Backs Deep-Sea Pipeline to India', *The Hindu*, 17 December 2013.
126 Turkmenistan supplies gas to Iran for its internal use. Earlier Iran had announced that it may not need to continue importing Turkmen gas. See Chapter 3.
127 Ankit Panda, 'India, Iran and Oman Open Talks on Deep Sea Gas Pipeline', 1 March 2014. http://thediplomat.com/2014/03/india-iran-and-oman-open-talks-on-deep-sea-gas-pipeline/. Accessed on 31 March 2014. See Also Bagchi, n. 117.
128 See Chapter 3 for more on this.
129 *Oman Daily Observer*, 19 February 2015.
130 Jung, n. 7, p. 78. After eight years of studies and about Rs. 330 crores spent, the Petroleum Ministry formally closed the $1.8 billion Oman-India pipeline proposal on 22 September 2002. *The Hindu*, 23 September 2002.
131 'Pakistan's Continental Shelf Extension Programme', *National Institute of Oceanography, Ministry of Science and Technology, Government of Pakistan*, 10 March 2015. http://www.niopk.gov.pk/cse.html. Accessed on 10 June 2015.
132 In 1995, Pakistan blocked a proposed deep sea line form Oman to India through its EEZ. Amitav Ranjan, 'Stuck in the Pipeline: A $4 billion Deep Sea Gas Project', *Indian Express*, 13 May 2015.
133 'Muscat Agreement on the Delimitation of the Maritime Boundary between the Sultanate of Oman and the Islamic Republic of Pakistan', 12 June 2000. http://www.un.org/depts/los/LEGISLATIONANDTREATIES/PDFFILES/TREATIES/OMN-PAK2000MB.PDF. Accessed on 13 May 2015.

CONCLUSIONS
Legacy, leads and lessons

Hurree Babu instructs the young hero in the mysteries of the Great Game. 'You sit tight, Master O'Hara . . . It concerns the pedigree of a white stallion.'
'Still? That was finished long ago.'
'When everyone is dead, the Great Game is finished. Not before.'[1]
[T]he oil industry lobby . . . comprises a rapacious welter of numerous think-tanks, law firms, investment bankers, trade associations, pipeline construction companies, aspiring journalists, television talking heads, big oil-controlled politicians, ambitious academics, hungry local officials, our agile expatriates there and unsettled Caspian *emegres* here – all united in the desire to slice off a pecuniary piece from the anticipated huge investment pie in the contracts, assignments and consulting fees.[2]

The quotes draw attention to an old and a recently revived phrase the 'Great Game' and identify a plethora of interested players seeking to make a fortune out of it. The second quote refers to oil, but it is just as applicable to gas. The concluding chapter seeks to bring the study under an overarching umbrella of framework before embarking on the journey of picking up the leads from the previous chapters and daring to peep into the future and venture into the territory of prescription.

The great gas game

The Great Game has been an overused, misused and sometimes abused term. It remains a useful trope, nonetheless. The policy wonks prescribe it, the powerful play it and the people understand it. Great Britain and Russia cast the first dice in the game and gave it a common name – the Great Game in English and Bolshaya Igra in Russian.

The vast chessboard on which the struggle for empire took place between London and Moscow stretched from the Caucasus in the West along the great deserts and mountain ranges to Tibet in the East. The Great Game, as it has come to be called, involved espionage and adventure in equal measure. The end result of the game was a historic compromise between them, under which the Russians were given a free hand in Central Asia, and India remained a jewel in the British Crown. The Great Game was not all about heartbreaking defeats, breathtaking suspense and exhilarating victories. On the ground, it was death and destruction. And, fun and play were not for the pawns.

The British Empire since then has vanished and the Russian Empire, after 70 years of Soviet Union, has vanished as well. The term has been resurrected and utilised with a vengeance after the Cold War. There are some striking similarities and some undeniable differences. So, is the analogy between the old and the new game totally incorrect?

The principal players of the old game were Britain and Russia. Today, the baton is held by the US, Russia and China. Although the game is played on the same field, the trophy is different. In olden times, the British played in Central Asia with an eye on India; today the energy potential of the region has made it a prize in its own right. The earlier game was played in the terrain of high politics – imperial ambitions dictating decisions of war and peace. Today, it is played in the territory of low politics – economic aims seeking uninterrupted energy at reasonable price. Is the high politics/low politics dichotomy sustainable? Or, has the pursuit of energy moved to the arena of high politics?

The old gamers occasionally forged military alliances to withstand the enemy better. The new gamers, by that logic, should forge economic alliances, that is cartels – the producers and exporters on one side and the consumers and importers on the other. In the gas game, there are no clear fault-lines. One can juggle with many permutations and combinations to classify the gas gamers. A producer can be a transit state and a consumer – like Russia. A consumer can be a transit state – like in most cases. And a consumer can also be a producer – like in many cases.

In spite of this tapestry, the countries that are mostly producers and the countries that are mostly consumers have sought to come together. The Gas Exporting Countries Forum (GECF) and the Gas Buyers Club (GBC). Both have stopped short of crossing the Rubicon to emerge as cartels. The gas deals are long-term bilateral contracts with a take-or-pay clause. (Strangely, there is no deliver-or-pay the compensation clause in the gas contracts!) This means the buyers are obligated to pay for the gas whether they take it or not. The rigidity of bilaterally agreed to gas prices leave little room for the GECF to influence gas production and prices. A cartel would necessarily require a swing producer to monitor and adjust the production and price – like the Saudis were doing till recently within the OPEC. Russia, the largest gas exporter would rather go it alone. Will the US, in the aftermath of the shale bonanza, agree to play that role?

The Gas Buyers Club has remained just that; a club. China, the largest importer would rather go its own way leaving the club toothless. A mention needs to be made of Putin's failed initiative to form a 'Gas Alliance' with the three gas-rich Central Asia states – Kazakhstan, Turkmenistan and Uzbekistan. It was immediately rejected as the Central Asians wanted Russia to lower the transit fee for their gas traversing through its territory to Europe, instead.

What then of the pawns of the old game? They were helpless sufferers or irrelevant bystanders. In their current reincarnation, they have evolved into gas-rich independent sovereign states. Their reach is limited and their movement is slow. They, nonetheless, are players in their own right. Creating their own space, looking out for good opportunities and right timings. Turkmenistan rejected Putin's idea of forming a Gas Alliance and accused Russia for the explosion in its pipeline. At the same time, its assertive behaviour has not crossed the red line; it has pulled back from the West-sponsored pipeline proposals.

Staying with the Great Game but moving on from the gas assets to the gas pipelines. The task of laying the pipelines involves sheer physical work, immense financial commitment and the most complex technological input. A gas pipeline is the result of a specific agreement between two or more parties. It is a reliable indicator of the relative power balance between the parties, a measure of their need to sell/buy the gas, and a measure of their willingness to compromise. A gas pipeline is likened to the railway lines in the nineteenth century as both facilitate transport and reflect political preferences of the powerful.

Like any cross-border trade, the gas pipelines benefit the exporting, transit and importing countries. A pipeline necessitates infrastructure development in border areas, which are usually remote, in difficult terrain and away from important cities and mainstream economy. It stimulates economic activities, generates employment in the pipeline construction and in collateral development projects. The project companies routinely undertake development of the pipeline neighbourhoods like providing employment, electricity, water, health services, schools, roads, parks and so on.

Just like McDonalds in two neighbouring countries do not necessarily lead to no-war situation between them; so doesn't a pipeline running through two or more states necessarily lead to improved political relations or generate peace between them. It does, at times, make strange bedfellows. Consider this. For decades the Russians have suspected Turkey of supporting the Chechens – their most dreaded violent separatists. Turkey has been harbouring dark suspicions that Russia extends asylum and protection to the Partiya Karkeren Kurdistan (PKK, Kurdistan Workers Party) that has been waging a war on the state of Turkey. The two are now busy planning and constructing a long pipeline to carry Russian gas to Turkey and beyond. The process in the aftermath of Turkey shooting down a Russian warplane in late 2015 has only led to its 'suspension' and not cancellation. Similarly, Turkey eagerly built a pipeline to Greece – its enemy across the Mediterranean – to pass on some of its surplus gas in order not to have to pay penalty to Russia and Iran under the take-or-pay clause. It was a milestone in the long period of distrust between the two. In 2005, the Turkish prime minister Recep Tayyip Erdogan and the Greek prime minister Kostas Karamanlis went on to jointly lay the foundation of the Interconnector Turkey–Greece–Italy pipeline.

At times, a pipeline may lead to hard bargains, threats, disputes and conflicts resulting in interruptions in the flow of gas. The longer the length of a pipeline and greater the number of countries through which it passes, the stronger the possibility of interruptions. A gas supplier may use the gas as a potent economic weapon to bring the importing country to a standstill. Russia has stopped the supply to Ukraine and Georgia several times. A transit country – most of the times also an importing one – would demand certain amount of gas for itself. It would also seek to constantly raise its fees for allowing gas passage through its territory. Or, the gas may simply be pilfered during the passage through transit, like the Russian gas in Ukraine.

A transit country – sometimes also an exporting country – would refuse permission to build a line, create obstruction during construction or after completion and seek to reduce the flow to protect its own share of the market. The Saudi objection to the Qatari gas line to the United Arab Emirates (UAE) and Oman is a case in point. In the mid-1990s Gazprom, as the sole owner of the pipeline that took Turkmen gas to Europe via Russia, halted Turkmen exports and demanded an increase in transit fees. Turkmenistan had little choice but to agree, as efforts to build viable export routes via Afghanistan and Iran had largely failed. There were rumours that the diameter of the pipe from Iran to Armenia was reduced from 1,420 to 700 millimetres under pressure from Gazprom, so that Iran could not export its gas to Europe and compete with the Russian share of the European market.

The exporting, transiting and importing countries may not remain the same forever. The US, the largest importer of energy, is all set to become a credible exporter. In the most dramatic swap of roles, Egypt that used to supply roughly 40 per cent of Israel's gas imports will now be importing gas from the newly discovered gas fields in Israel.

A pipeline may get enmeshed into several other bilateral issues. For example, the Russia–Ukraine relations always involved a package of agreements including the future of Russian Black Sea Fleet and the Ukrainian nuclear weapons together with the pipeline issue. A line could become a target in a war fought on different grounds. During the Russia–Georgia war in 2008, a Russian bomb fell very close to the Baku–Tbilisi–Ceyhan oil and Baku–Tbilisi–Erzurum gas pipelines. Many Western analysts suggested that one reason Russian aviators dropped a bomb only 50 metres from the pipelines was to highlight the insecurity of the link between the Caspian and the West as it bypassed Russian territory. The counter-argument held that if Russia wanted to bomb the lines, it would have kept trying until it hit the target rather than giving up after one attempt; also it was not in Russia's interest to bomb the pipelines as doing so would have badly alienated Azerbaijan.

A regime change or an attempted regime change is often explained with reference to a gas pipeline. The Russia–Georgia war in 2008 can be traced back several years before the actual conflict. There were two cut-offs of Russian gas to Georgia in 2001. The Russian gas company Itera had already bought 50 per cent of the Georgian gas distributor Gruzgas and was in the process of purchasing the remaining 50 per cent. The country became the first major theatre of confrontation between the US and Russia and the first country to witness a Colour Revolution. The 'Rose Revolution' overthrew Eduard Shevardnadze and brought in Mikheil Saakashvili as the president.

A second attempt at regime change is still going on in Syria. The country is a key link in two rival pipeline projects: Iran–Iraq–Syria–Lebanon on to Greece and the Qatar–Saudi Arabia–Jordan–Syria–Turkey. Whether the Assad regime survives or a change of regime happens there would determine the global gas system in a large way. The line originating in Qatar would pry Turkey loose from dependence on Iranian supplies, it would severely curtail the Russian near monopoly as the gas supplier to Europe and it would facilitate Israel's gas export to Europe. There are many interesting conspiracy theories making rounds around these rival pipeline proposals. One sees a clear connection in the timing between the signing of the memorandum on the Iran–Iraq–Syria–Lebanon gas pipeline and the beginning of violent uprising in Syria. The other notes a coincidence between the fiercest fighting in places inside Syria and the proposed route of the Qatari pipeline through its territory. The Saudi Prince Bandar was reported to have threatened Putin to let the Chechen rebels loose during the Sochi Olympics, if Russia did not go along with the regime change in

Syria that would have made the Qatar–Saudi Arabia–Jordan–Syria–Turkey line possible.

Ukraine has seen regime changes many times over. In 2004–2005, the 'Orange Revolution' was the second Colour Revolution after Georgia that resulted in a re-vote after massive protests against the elections that were seen to be rigged in favour of the victory of Viktor Yanukovych over his contestant Viktor Yushchenko. The re-vote resulted in favour of the latter. Yanukovych and Yushchenko were the favourite candidates of Russia and the US respectively. The Russian diplomacy to retain control over its former space and the US diplomacy to extend its control over the same were bound to come to a collision. Ukraine conflict is much more than a struggle over the pipeline. In a classic case of competitive geopolitics between Russia and the West, the country is in a state of civil war and virtually divided down the middle.

A reference was made to geography in the introductory chapter. Geography is not all about rivers and mountains. It informs foreign policy and military doctrines in a big way. And geography is eternal; the most profound changes in regime or national borders or the state ideology does not have any influence on the geography. A perfect example of geography dictating pipeline routes is the complex repertoire of moves and gestures accompanying the construction of the Baku–Tbilisi–Ceyhan oil pipeline and the parallel Baku–Tbilisi–Erzurum gas pipeline. The US consistently championed the route to Turkey, even though it was the least feasible of options. And it equally consistently blocked the route via Iran, even though it was the shortest, safest and the cheapest of the options. The route to Turkey involved the Azeri gas from Baku moving through Armenia. Since Armenia accuses Turkey of genocide of the Armenians in 2015, that route had to be ruled out. The twists and turns finally led to the Baku–Tbilisi–Erzurum route avoiding Armenia. The pipeline in Georgia had to be buried close to Armenia and in Turkey close to the Kurdish area. The war in Chechnya, trouble in Dagestan, conflict over Nagorno-Karabakh, separatist movements in Abkhazia and Ossetia: all posed their own political problems. Again, the line in Georgia had to go through the mountains at an altitude of 3 kilometres and close to the country's famous mineral springs of Borjomi. The environment NGOs were very sensitive and active to protect the spring. In short, the successful laying of the two lines was a technological as well as a diplomatic accomplishment. And the geography was a decisive variable every step of the route.

A few projections into the future are in order, even at the risk of most of them not coming through or only partially coming through. The shale gas has arrived. The US will be much ahead of the rest in the twin technology of horizontal drilling and fracking. China is estimated to have larger deposits of shale than the US, but it faces insurmountable obstacles: technological deficit, population density and availability of land and water. Europe will have to overcome similar obstacles in addition to legal restraints and public opinion that is hostile to shale extraction on environmental grounds. Russia, Iran and Qatar, the possessors of vast gas reserves, will lose out to the new arrival.

The world of energy tomorrow will be defined by an energy self-sufficient US and an assertive energy thirsty China. Will the US be a bigger game changer on the supply side or China on the demand side? The future scenario will also see a fundamental shift in the major importers from Europe to Asia where there is a rising demand for energy and in the major exporters from the Caspian to the US that is debating relaxation in its gas exports policies.

The future will be witness to a steady march of technological advance that will constantly make the 30 per cent global gas that is orphan today into proved reserves. It may even make the fossil energy redundant as new sources are discovered and harnessed. On the flip side, the world may be reduced to a helpless bystander to occasional/increasing acts of sabotage and terrorism.

Global warming and climate change will have an increasing impact on gas extraction and transport. The global warming will prove to be a boon to Russia. Extracting energy in the forbidding landscape of the Arctic would be a less arduous task. The ice breaker ships may not be required to reach the energy under deep freeze. On the other hand, environment consideration has resulted in stalling or rerouting the energy passage. For instance, Russia insists on a ban on pipelines along the bottom to be included in Caspian Sea's legal status on environment ground. Turkey insists on double-hull tankers, higher insurance premiums and a right to regulate traffic in Bosporus along Istanbul to protect the fragile Strait. The EU Parliamentary Panel has responded to the environmental associations in its member states and held that a Baltic passage must be approved by all countries with a Baltic coastline.

The recent discoveries in the eastern Mediterranean will facilitate the entry of new members into the exclusive group of gas exporters, shifting the gas landscape further west. The gas trade linkages between Israel, Egypt and Jordan and a possible dispute over gas fields between Israel and Lebanon are set to create their own dynamics in an already unstable region.

Sanctions on Iran are unravelling, even if they are not formally lifted or only staggered. The Nabucco project, long in hibernation, may yet be resurrected with an eager Iran waiting to join it.

And finally, the future gas network will not be a stand-alone project, but will be an important component in the competing grand visions. Today, the movement is towards a broader connectivity that would subsume the gas pipelines into its gamut. The International North-South Transport Corridor (INSTC) between India, Iran and Russia was one of the first broad-based attempts at connectivity, even though it was not about the pipelines. The US Silk Road Strategy was all about the energy pipelines with an improbable goal of securing the centre stage for Afghanistan. It was an exercise in redesigning the INSTC geometry by turning the north–south line a 90-degree westward. The north–south was to become the east–west link with the sole aim of keeping Russia and Iran out of the global energy map. The dual exclusion would have restricted the choice of the buyers in Europe and also the choice of the sellers in West and Central Asia. The energy companies were equally reluctant to pursue strategy rather than profit in the pipeline choices.

Europe has its own projects at connecting to the Caspian. Unlike the US Silk Road project, they are less grand in their conception and are commerce-driven. The Southern Gas Corridor will arrange to coordinate a number of pipelines to reach Europe with the Caspian gas. The Transport Corridor Europe–Caucasus–Asia (TRACECA) focuses on marine transport, air routes, roads and rails, transport infrastructure and security. The INOGATE has been an energy technical assistance programme between the European Union and 11 Partner Countries in East Europe, Caucasus and Central Asia.

The Shanghai Cooperation Organisation (SCO), a multilateral group in Eurasia and constituting China, Russia, Kazakhstan, Kyrgyzstan, Tajikistan and Uzbekistan, is committed to prioritise joint energy projects including oil and gas sectors and exploration of new hydrocarbon reserves.

The Brazil–Russia–India–China–South Africa (BRICS) representing roughly 40 per cent of the world population and engaging with the broad-based agenda towards democratising international relations does not explicitly mention energy cooperation, except to acknowledge fossil fuel as one of the major sources of energy while committing itself to encourage renewable and clean energy.

Russia is committed to establishing the Eurasian Economic Union, whose membership to date remains confined to Russia, Kazakhstan, Belarus and Armenia. It is a single market and provides for common transport and energy policies.

Turkey has formed an 'International Forum on the Role of Customs Administration on Facilitating and Promoting Trade among Silk Road Countries' in Antalya, which aims to take concrete steps regarding simplification of border crossing procedures and trade facilitation among the Silk Road countries.

Surpassing these initiatives is the grandiose Chinese 'One Belt, One Road' project. It is an overland as well as maritime project that goes much beyond the Caspian or even Eurasia. It extends to the entire Eastern Hemisphere, which is set to recover its historical position at the centre of the world and, probably, China at the very epicentre of it. The project is backed up by an equally ambitious and expansive Asian Infrastructure Investment Bank that is mandated to provide finance for infrastructure development towards economic betterment and integration. It is an impressive project in its size and scope.

India has proposed a transnational initiative meant to revive its ancient maritime routes and cultural linkages with countries in the region. Titled 'Project Mausam: Maritime Routes and Cultural Landscapes across the Indian Ocean', it focuses on the natural wind phenomenon, especially monsoon winds used by Indian sailors in ancient times for maritime trade, that has shaped interactions between countries and communities connected by the Indian Ocean. It is more a cultural construct rather than a political–strategic one.

Moving on from projection into the future to the prescription for the present. India is surrounded by gas-rich countries: Myanmar and Bangladesh to the east, the Gulf to the west and Central Asia to the north. Unfortunately, it has no external gas pipeline coming in, going out or transiting through its territory as yet. Since it is set on a high economic growth with a scarcity of indigenous supply of hydrocarbons, energy security constitutes a high quotient in its overall security calculus. Exploring for gas onshore and offshore, diversifying sources of supply, acquiring latest technology, mobilising finances, forging alliances at the level of the companies as well as governments, promoting energy efficiency and conservation – all these have been oft-repeated prescriptions. And all of them deserve constant and committed pursuit.

Out of the previous suggestions, diversification needs to be reiterated. It must be made the topmost priority that India has to pursue. There are security and price constraints about the IPI pipeline; there has been a Chinese checkmate with the Myanmar and Bangladesh supplies; and the TAPI will take time. Is the LNG, therefore, the fuel of choice? As noted, there are technological advances on the piped and LNG deliveries, and there are risks of sabotage to both. It is difficult to predict whether the piped gas would benefit more from the technological innovations or the LNG. Equally difficult is to predict the occurrence and destructiveness of the threats to the LNG or the piped gas. Therefore, the sources of supply and also the form in which the gas is imported need to be diverse as well. India will need imported gas in increasing amounts; therefore, there is no either-or. We need the piped gas as well as the LNG from diverse sources.

References have been made throughout the study to the role of geography. It is eternal, immutable. The most profound changes in economic, political or ideological transformations will not affect geography. The IPI has remained in hibernation for decades – due to sanctions on Iran, but also because the line would have to transit through Pakistan with its inherent security concerns. After the recent extension of Pakistan's continental shelf, the Middle East to India Deep Water Pipeline (MEIDP), an initiative by the SAGE, would also have to traverse the Pakistan EEZ. In this situation, either permission from Pakistan or inclusion of Pakistan in the project is the only way forward. If Pakistan is a partner in the TAPI project; so could it be in the IPI and the MEIDP. A country the size, population, economy and aspirations like India has no other choice but diplomacy and accommodation, with an adequate portion of caution in the mix.

To conclude, the game will continue long after the white stallion is dead. And the game is definitely not confined to gas. It is the states relentlessly pursuing their national interests – singly, bilaterally, multilaterally; in cooperation, competition, conflict. If one takes defensive realism as the broad context within which to understand and explain the game, then an increasing number of state players, influence of domestic circumstances and acknowledgement of non-state actors would have to be taken on board within this theoretical construct. Purity of the theory and the reality on the ground cannot be mutually exclusive.

Notes

1 Rudyard Kipling, *Kim* (Macmillan, London, 1944), p. 316.
2 Alec Rasizade, 'The Caspian Energy Legend and the "Great Game" of Concomitant Pipelines', *Iranian Journal of International Affairs*, Vol. 14, nos 1–2, Spring–Summer 2002, p. 27.

BIBLIOGRAPHY

Primary sources

The Constitution of the Islamic Republic of Iran (Islamic Propaganda Organisation, Tehran, 1990).

EIA, 'Shale Oil and Gas Resources Are Globally Abundant', 2 January 2014. http://www.eia.gov/todayinenergy/detail.cfm?id=14431

Hiyashi, Nobuyuki, 'Natural Gas in China: Market Evolution and Strategy', *International Energy Agency Working Papers*, 2009.

http://indiabudget.nic.in (Ministry of Finance, Government of India).

http://pogc.ir (Pars Oil and Gas Company).

http://powermin.nic.in (Ministry of Power, Government of India).

http://www.gecf.org/Resource/GECF-History-File.pdf (The Gas Exporting Countries Forum).

Hydrocarbon Vision, 2025. www.indiaworldenergy.org

'IEA Natural Gas Security Study', *International Energy Agency*, 1995.

Integrated Energy Policy: Report of the Expert Committee (Planning Commission, Government of India, 2006). http://planningcommission.nic.in/reports/genrep/rep_intengy.pdf

Mark, Leonard and Nicu Popescu, 'A Power Audit of EU-Russia Relations', *Report by the European Council on Foreign Relations*, November 2007.

Medium Term Gas Market Reform, 'Energy Information Administration', 20 June 2013. http://iea.org/newsroomandevents/pressreleases/2013/june/name,39014,en.html

Medium Term Oil and Gas Markets, 2011, 'International Energy Agency'.

'Muscat Agreement on the Delimitation of the Maritime Boundary between the Sultanate of Oman and the Islamic Republic of Pakistan', *United Nations*, 12 June 2000. http://www.un.org/depts/los/LEGISLATION ANDTREATIES/PDFFILES/TREATIES/OMN-PAK2000MB.PDF

Natural Gas Market Review 2006: Towards a Global Gas Market (International Energy Agency, Paris), 2006.

Natural Gas Market Review, 2007 (International Energy Agency, Paris). http://www.iea.org/publications/freepublications/publication/gasmarket2007.pdf

'Pakistan's Continental Shelf Extension Programme', *National Institute of Oceanography, Ministry of Science and Technology, Government of Pakistan*, 10 March 2015. http://www.niopk.gov.pk/cse.html

Putin, Vladimir, 'A New Integration Project for Eurasia: The Future in the Making', *Izvestia* (Moscow), 3 October 2011. Reproduced in 'The Permanent Mission of the Russian Federation to European Union'. http://www.russianmission.eu/en/news/article-prime-minister-vladimir-putin-new-integration-project-eurasia-future-making-izvestia-3-

Ralf, Dickel, 'Cross-Border Oil and Gas Pipeline Projects: Analysis and Case Studies', *World Bank Review Version*, 5 September 2001.

Removing Obstacles to Cross-Border Oil and Gas Pipelines: Problems and Prospects, 'Energy Sector Management Assistance Program (ESMAP)', UNDP/World Bank, Washington, DC, 2003.

'Rising Powers: The Changing Geopolitical Landscape', *Report of the National Intelligence Council's 2020 Report*. http://www.globalsecurity.org/intell/library/reports/2005/nic_globaltrends2020_s2.htm

Scherbak, Yuri, Ambassador Extraordinary and Plenipotentiary of Ukraine, at the Plenary Session of the Days of Science, National University of Kyiv Mohyla Academy, Kyiv, Ukraine, January 30, 2004. Published by the 'UKRAINE REPORT', 2004, No. 20, February 4, 2004. http://www.artukraine.com/buildukraine/shcherbak3.htm, accessed 25 January 2013.

www.bp.com (British Petroleum).

www.dghindia.org (Directorate General of Hydrocarbons, Ministry of Petroleum and Natural Gas, India).

www.eia.doe.gov (US Energy Information Administration).

www.eni.it (Eni S P A).

www.europeanenergy.com (EU Official Website on Energy Issues).

www.iea.org (International Energy Agency).

www.iocl.com (Indian Oil Corporation Limited).

www.meaindia.nic.in (Ministry of External Affairs, Government of India).

www.mfa.gov.ir (Ministry of Foreign Affairs, Government of Iran).

www.mofa.gov.pk (Ministry of Foreign Affairs, Government of Pakistan).

www.oil-india.com (Oil India Limited).

www.ongcindia.com (Oil and Natural Gas Corporation, India).

www.pngrb.gov.in (Petroleum and Natural Gas Regulatory Board, India).

www.usgs.gov (US Geological Survey).

Books

Abu-Lughod, Janet L., *Before European Hegemony: The World System AD 1250–1350* (Oxford University Press, New York, 1989).

Alam, Anwar, ed., *India and West Asia in the Era of Globalisation* (New Century Publications, New Delhi, 2008).

Alam, Anwar, ed., *India and Iran: An Assessment of Contemporary Relations* (New Century Publications, New Delhi, 2011).

Ali, Salim H., *Energising Peace: The Role of Pipelines in Regional Security* (The Brookings Doha Centre Analysis), no. 2, July 2010. http://www.brookings. edu/research/papers/2010/07/middle-east-ali

Anceschi, Luca, *Turkmenistan's Foreign Policy: Positive Neutrality and the Consolidation of the Turkmen Regime* (Routledge, London, 2008).

Atlantic Council and the Central Asia-Caucasus Institute at the Johns Hopkins University's School of Advanced International Studies, *Strategic Assessment of Central Eurasia*. http://www.acus.org/Publications/policypapers/inter nationalsecurity/Central% 20Eurasia.pdf

Audinet, Pierre, P. R. Shukla and Frederic Grare, eds., *India's Energy: Essays on Sustainable Development* (Manohar, New Delhi, 2000).

Bandey, Aijaz A., ed., *Silk Route and Eurasia: Peace and Cooperation* (University of Kashmir Press, Srinagar, 2011).

Bellacqua, James, ed., *The Future of China-Russia Relations* (University Press of Kentucky, Lexington, 2010).

Bhaskar, Bala, *Energy Security and Economic Development in India: A Holistic Approach* (TERI, New Delhi, 2013).

Bhatia, Rajiv K., Vijay Sakhuja and Indrani Talukdar, *India and Russia: Deepening the Strategic Partnership* (Shipra, New Delhi, 2014).

Bose, Shrinjoy, *Energy Politics: India–Bangladesh–Myanmar Relations* (Special Report no. 45, IPCS, New Delhi, July 2007).

Brzezinski, Zbigniev, *The Grand Chessboard: American Primacy and Its Geostrategic Imperatives* (Basic Books, New York, 1997).

Busby, Rebecca L., ed., *Natural Gas in Nontechnical Language* (Institute of Gas Technology, PennWell, Tulsa, 1999).

Calder, Kent A., *Asia's Deadly Triangle* (Nicholas Brealey Publishing, London, 1997).

Chandra, Vivek, *Fundamentals of Natural Gas: An International Perspective* (PennWell Corporation, Tulsa, 2006).

Cheterian, Vicken, *Dialectics of Ethnic Conflicts and Oil Projects in the Caucasus* (Programme for Strategic and International Security Studies, The Graduate School of International Studies, Geneva, 1997).

Cohen, Saul Bernard, *Geopolitics of the World System* (Rowman and Little Field, Lanham, 2003).

Cole, Bernard D., *Sea Lanes and Pipelines: Energy Security in Asia* (Praeger Security International, Westport, 2008).

Colgan, Jeff D., *Petro-Aggression: When Oil Causes War* (Cambridge University Press, Cambridge, 2013).

Collins, Gabriel B., Andrew S. Erickson, Lyle J. Goldstein and William S. Murray, eds., *China's Energy Strategy: The Impact on Beijing's Maritime Policies* (China Maritime Studies Institute and Naval Institute Press, Annapolis, 2008).

Cordner, Lee, *Offshore Oil and Gas Safety and Security in the Asia Pacific: The Need for Regional Approaches to Managing Risks* (S. Rajaratnam School of International Studies, Monograph number 26, 2013).

Crandall, Maureen S., *Energy, Economics and Politics in the Caspian Region: Dreams and Realities* (Praeger Security International, Westport, 2006).

Dadwal, Shebonti Ray, *Rethinking Energy Security in India* (Knowledge World, New Delhi, n.d.).

Dadwal, Shebonti Ray, *The Geopolitics of America's Energy Independence: Implications for China, India and the Global Energy Markets* (Institute for Defence Studies and Analyses, New Delhi, 2013).

Dellecker, Adrian and Thomas Gomart, eds., *Russian Energy Security and Foreign Policy* (Routledge, New York, 2011).

Devre, Sudhir T., ed., *A New Energy Frontier: The Bay of Bengal Region* (Institute of Southeast Asian Studies, Singapore, 2008).

Dodds, Klaus, *Global Geopolitics: A Critical Introduction* (Pearson Education Ltd, Essex, 2005).

Doraiswamy, Rashmi, ed., *Energy Security: India, Central Asia and the Neighbourhood* (Manas Publications, New Delhi, 2013).

Duyne, Dorothy van, *The Straits of Malacca: Strategic Considerations* (US Naval Academy, Annapolis, 2007).

East, Gordon W., *The Geography behind History* (Norton, New York, 1965).

Ebenback, Ben W., *Energy Resources: Availability, Use and Impact* (PennWell, Tulsa, 1995).

Ebinger, Charles K., *Energy and Security in South Asia: Cooperation or Conflict?* (Brookings Institute Press, Foundation Books, Delhi, 2013).

Emirates Center for Strategic Studies and Research, *Gulf Energy and the World: Challenges and Threats* (Abu Dhabi, 1997).

Emirates Center for Strategic Studies and Research, *Asian Energy Markets: Dynamics and Trends* (Abu Dhabi, 2004).

Emirates Center for Strategic Studies and Research, *The Gulf Oil and Gas Sector: Potential and Constraints* (Abu Dhabi, 2006).

Emirates Center for Strategic Studies and Research, *Gulf Oil and Gas: Ensuring Economic Security* (Abu Dhabi, 2007).

Engdahl, William, *A Century of War: Anglo-American Oil Politics and the New World Order* (Pluto Press, London, 2004), Revised Edition.

Escobar, Pepe, *Globalistan: How the Globalised World Is Dissolving into Liquid War* (Nimble Books, Ann Arbor, 2007).

Escobar, Pepe, *Obama Does Globalistan* (Nimble Books, Ann Arbor, 2009).

Freeman, Donald B., *The Straits of Malacca: Gateway or Gauntlet?* (McGill-Queen's University Press, 2003).

Gall, Carlotta and Thomas de Waal, *Chechnya: A Small Victorious War* (MacMillan, London, 1997).

Gallick, Edward C., *Competition in the Natural Gas Pipeline Industry: An Economic Policy Analysis* (Praeger, Westport, 1993).

Gray, Collin S., *Geopolitics, Geography and Strategy* (Frank Cass, London, 1999).

Grygiel, Jakub J., *Great Powers and Geopolitical Change* (John Hopkins University Press, Baltimore, 2006).

Guo, Boyun, Shanhong Song, Ali Ghalambor and Tian Lin, *Offshore Pipelines: Design, Installation and Maintenance* (Gulf Professional Publishing, Elsevier, 2013).

Gupta, Ranjit, Abu Backer Bagader, Talmiz Ahmad and N. Janardhan, eds., *India and the Gulf: What Next?* (Gulf Research Centre, Cambridge, 2013).

Harris, Katherine T., ed., *Geopolitics of Oil* (Nova Science Publishers, New York, 2009).

Hiro, Dilip, *Blood of the Earth: The Battle for the World's Vanishing Oil Resources* (Penguin Books, New Delhi, 2008).

Holms, James R., Andrew C. Winner and Toshi Yoshihara, *Indian Naval Strategy in the 21st Century* (Routledge, London, 2009).

Jackson, Brian A., Lloyd Dixon and Victoria A. Greenfield, *Economically Targeted Terrorism: A Review of the Literature and a Framework for Considering Defensive Approaches* (Rand Corporation, Santa Monica, 2007).

Jensen, James T., 'The Future of Gas Transportation in the Middle East Region: LNG, GTL and Pipelines', in *The Gulf Oil and Gas Sector: Potential and Constraints* (Emirates Center for Strategic Studies and Research, Abu Dhabi, 2006).

Kaplan, Robert D., *The End of the Earth* (Random House, New York, 1996).

Kaplan, Robert D., *Hog Pilots, Blue Water Grunts: The American Military in the Air, at Sea, and on the Ground* (Random House, New York, 2007).

Kaplan, Robert D., *The Return of History and the End of Dreams* (Atlantic Books, London, 2008).

Kaplan, Robert D., *Monsoon: The Indian Ocean and the Future of American Power* (Random House, New York, 2010).

Karl, Terry Lynn, *The Paradox of Plenty: Oil Booms and Petro-States* (University of California Press, 1997).

Kaw, Mushtaq A. and Aijaz A. Banday, eds., *Central Asia: Introspection* (Centre of Central Asian Studies, Srinagar, 2006).

Kemp, Geoffrey and Robert E. Harkavy, *Strategic Geography and the Changing Middle East* (Brookings Institution Press, Washington, DC, 1997).

Kennedy, John L., *Oil and Gas Pipeline Fundamentals* (PennWell Publishing Company, Tulsa, Oklahoma, 1993), 2nd Edition.

Kennedy, Paul, *The Rise and Fall of the Great Powers: Economic Change and Military Conflict from 1500 to 2000* (Random House, New York, 1987).

Keohane, Robert O. and Joseph S. Nye, *Power and Interdependence: World Politics in Transition* (Little Brown and Company, Boston, 1977).

Khanna, Parag, *The Second World: Empires and Influence in the New Global Order* (Allen Lane, London, 2008).

Khosla, I. P., ed., *Energy and Diplomacy* (Konark Publishers, New Delhi, 2005).

Klare, M. T., *Resource Wars: The New Landscape of Global Conflict* (Henry Holt and Company, New York, 2001).

Kurlantzik, Joshua, *Charm Offensive: How China's Soft Power Is Transforming the World* (Yale University Press, New Haven, 2007).

Leghari, Faryal, ed., *Gulf-Pakistan Strategic Relations* (Gulf Research Centre, Dubai, 2008).

MacDonald, Juli A., Amy Donahue and Bethany Danyluk, *Energy Future in Asia: Final Report* (Booz Allen Hamilton, McLean, November 2004).

Malakar, S. N., ed., *India's Energy Security and the Gulf* (Academic Excellence, New Delhi, 2006).

Manning, Robert A., *The Asian Energy Factor: Myths and Dilemmas of Energy, Security and the Pacific Future* (Palgrave, New York, 2000).

Marketos, Thrassy N., *China's Energy Politics: The Shanghai Cooperation Organisation and Central Asia* (Routledge, London, 2009).

McAllister, E. W., *Pipeline Rules of Thumb Handbook: A Manual of Quick, Accurate Solutions to Everyday Pipeline Engineering Problems* (Gulf Professional Publishing, Elsevier, 2013).

McKillop, Andrew, *The Doomsday Machine* (Palgrave Macmillan, Basingstoke, 2012).

Menashri, David, *Central Asia Meets the Middle East* (Frank Cass, London, 1998).

Miyamoto, Akira, *Natural Gas in Central Asia: Industries, Markets and Export Options of Kazakhstan, Turkmenistan and Uzbekistan* (The Royal Institute of International Affairs, London, 1997).

Muni, S. D. and Girijesh Pant, *India's Search for Energy Security* (Rupa and Company, New Delhi, 2005).

Naumkin, Vitaly, ed., *Turkey between Europe and Asia* (Institute of Oriental Studies, Moscow, 2000).

Oil and Gas Exploration and Production: Reserves, Costs, Contracts (Centre for Economics and Management, Institut Francais du Petrole Publications, Paris, 2004).

Olcott, Martha Brill, *The Geopolitics of Natural Gas: Turkmenistan: Real Energy Giant or Eternal Potential?* (Center for Energy Studies, Rice University's Baker Institute and Belfer Center for Science and International Affairs of the Harvard Kennedy School, 10 December 2013).

On the Brink: Desparate Energy Pursuits in South Asia (Panos South Asia, n.p., 2006).

Paik, Keun Wook, *Gas and Oil in Northeast Asia: Policies, Projects and Prospects* (Royal Institute of International Affairs, London, 1995).

Paik, Keun-Wook, 'Tarim Basin Energy Development: Implications for Russian and Central Asian Oil and Gas Exports to China' (Royal Institute of International Affairs, *CACP Briefing*, no. 14, November 1997).

Patey, Luke, *The New Kings of Crude: China, India, and the Global Struggle for Oil in Sudan and South Sudan* (Hurst and Company, London, 2014).

Peimani, Hooman, *The Caspian Pipeline Dilemma: Political Games and Economic Losses* (Praeger, Westport, 2001).

Pumphrey, Caroline, ed., *Energy and Security Nexus: A Strategic Dilemma* (Strategic Studies Institute, Pennsylvania, 2012).

Rao, Vikram, *Shale Gas: The Promise and the Peril* (Research Triangle Institute Press, North Carolina, 2012).

Rashid, Ahmad, *Taliban: Militant Islam, Oil and Fundamentalism in Central Asia* (Yale University Press, New Haven, 2000).

Reddy, B. S. and Nathan H. S. K., *Emerging Energy Insecurity: The Indian Dimension* (India Development Report, Oxford University Press, New Delhi, 2011).

Roberts, John, *Caspian Pipelines* (Royal Institute of International Affairs, Former Soviet South Project, London, 1996).

Saeid, Mokhatab, William A. Poe and James G. Speight, *Handbook of Natural Gas Transmission and Processing* (Elsevier, Burlington, 2006).

Santhanam, K. and Ramakant Dwivedi, eds., *India and Central Asia: Advancing the Common Interest* (IDSA and Anamaya Publishers, New Delhi, 2004).

Shaffer, Brenda, *Partners in Need: The Strategic Relationship of Russia and Iran* (Washington Institute for Near East Policy, Washington, DC, 2001).

Shambaugh, David, *Modernising China's Military: Progress, Problems, Prospects* (University of California Press, Berkley, 2004).

Simpfendorfer, Ben, *The New Silk Road: How a Rising Arab World Is Turning Away from the West and Rediscovering China* (Palgrave Macmillan, London, 2009).

Singh, Jasjit, ed., *Asia's New Dawn: The Challenges to Peace and Security* (Institute for Defence Studies and Analyses, New Delhi, 2000).

Singh, Jasjit, ed., *Oil and Gas in India's Security* (Knowledge World, New Delhi, 2001).

Skagen, Ottar, *Caspian Gas* (Royal Institute of International Affairs, London, 1997).

Stern, Jonathan P., *Third Party Access in European Gas Industries: Regulation-driven or Market-led?* (Royal Institute of International Affairs, London, n.d.).

Stern, Jonathan P., *The Russian Natural Gas 'Bubble': Consequences for European Gas Markets* (The Royal Institute of International Affairs, London, 1995).

Stern, Jonathan P., *Gas Pipeline Co-operation between Political Adversaries: Examples from Europe* (The Royal Institute of International Affairs, London, 2005).

Stern, Jonathan P., ed., *Natural Gas in Asia: The Challenges of Growth in China, India, Japan and Korea* (Oxford University Press, New York, 2008).

Stevens, Paul, ed., *International Gas: Prospects and Trends* (Macmillan, Houndmills, 1986).

Strange, Susan, *The Retreat of the State: The Diffusion of Power in the World Economy* (Cambridge University Press, Cambridge, 1996).

Sullivan, Arthur and Steven M. Sheffrin, *Economics: Principles in Action* (Pearson Prentice Hall, New Jersey, 2003).

Susan, Shirk, *China: Fragile Superpower* (Oxford University Press, New York, 2008).

Thakurta, Paranjoy Guha with Subir Ghosh and Jyotirmoy Chaudhari, *Gas Wars: Crony Capitalism and the Ambanis* (Paranjoy Guha Thakurta Publishers, New Delhi, 2014).

Trenin, Dmitri V., *Getting Russia Right* (Carnegie Endowment of International Peace, Washington, DC, 2007).

Tussing, Arlon R. and Bob Tippee, *The Natural Gas Industry: Evolution, Structure and Economics* (PennWell Books, Tulsa, 1995).

Vaughn, B., *Bangladesh: Political and Strategic Developments and US Interests* (Congressional Research Service, R41194, Washington, DC, April 2010).

Victor, David G., Amy M. Jaffe and Mark H. Hayes, eds., *Natural Gas and Geopolitics: From 1970 to 2040* (Cambridge University Press, Cambridge, 2006).

Walker, Martin, *CHIMEA: The Emerging Hub of the Global Economy* (A. T. Kearney Report, Washington, DC, 2008).

Waltz, Kenneth, *Theory of International Relations* (Addison-Wesley, Reading, 1979).

Winrow, Gareth, *Turkey and the Caucasus: Domestic Interests and Security Concerns* (The Royal Institute of International Affairs, London, July 2000).

Wybrew-Bond, Ian and Jonathan Stern, eds., *Natural Gas in Asia: The Challenges of Growth in China, India, Japan and Korea* (Oxford University Press, Oxford, 2002).

Yergin, Daniel, *The Prize: The Epic Quest for Oil, Money and Power* (Simon and Schuster, New York, 1991).

Yergin, Daniel, *The Quest: Energy, Security and the Remaking of the Modern World* (Allen Lane, London, 2011).

Zakaria, Fareed, *The Post American World* (Allen Lane, London, 2008).

Newspapers, journals and magazines

Afghan Voices (Sydney)
Al Manar News (Beirut)
Asia Pacific Issues (Honolulu)
Atlantic Monthly (Washington, DC)
Boston Globe (Boston)
Brown Journal of World Affairs (Providence)
Bulletin of Atomic Scientists (Chicago)
Central Asia Survey (Routledge, London)
China and Eurasia Forum Quarterly (Washington, DC)
CHINAUSFOCUS (Hong Kong)
Christian Science Monitor (Boston)
Daily Times (Islamabad)
Economic Times (New Delhi)
Economy Watch (Web Daily)
Energy Policy (Elsevier)
Express Tribune (Karachi)
Financial Times (London)
Forbes (New York)
Foreign Affairs (New York)
GlobalAsia (Seoul)

Global Research (Montreal)
GSP Occasional Paper Series (New Delhi)
Herald (Karachi)
The Hindu (New Delhi)
International Affairs (London)
International Affairs (Moscow)
International Affairs Review (Washington, DC)
ISS Alerts (Paris)
Journal of Energy Policy (Elsevier, London)
Journal of Energy Security (Maryland)
Journal of Eurasian Affairs (Moscow)
Journal of International Law and Politics (New York)
Journal of the Indian Ocean Region (London)
National Interest (Washington, DC)
Naval War College Review (Newport)
New York Times (New York)
Orbis (Philadelphia)
Pacific and Asian Journal of Energy (New Delhi)
Petroleum Economist (London)
Petroleum Intelligence Weekly (New York)
Policy Review (Washington, DC)
Proceedings (Maryland)
Southern European and Black Sea Studies (London)
Strategic Analysis (New Delhi)
Survival (London)
Terrorism Monitor (Washington, DC)
Washington Quarterly (Washington, DC)
World Politics Review (Tampa)
Yemen Post (Sanaa)

Newspaper, journal and magazine articles

Abubakr, Khaled, 'Future Gas: From North, South or East? Africa and Middle East Perspective', *7th European Gas Conference*, Oslo, 4 June 2013.

Ahmed, Nafeez, 'Syria Intervention Plan Fueled by Oil Interests Not Chemical Weapon Concern', *The Guardian*, 30 August 2013.

Alam, Shah, 'Iran's Hydrocarbon Profile: Production, Trade and Trend', *Strategic Analysis*, Vol. 25, Number 1, April 2001, pp. 119–134.

'Alert Major Cyber Attack Aimed at Natural Gas Pipeline Companies', *Christian Science Monitor*, 5 May 2012.

Alexeev, Igor, 'Fracking Fantasies: Has The Shale Bubble Already Burst?', *Economy Watch*, 26 August 2013.

Al-Hasan, Husam H., Faisal A. Qari and Syed M. Badruddoza, 'Saudi Aramco Faces the Desert Pipeline Challenge', *Pipelines International*, September 2013.

http://pipelinesinternational.com/news/saudi_aramco_faces_ the_desert_pipeline_challenge/083147/

Ali, Salim H., 'Energising Peace: The Role of Pipelines in Regional Security', *The Brookings Doha Centre Analysis*, Number 2, July 2010.

Aminuddian, Usman, 'Opportunities in the Development of the Oil and Gas Sector in South Asian Region', *Islamabad Papers*, Number 21 (Institute of Strategic Studies, Islamabad, September 2004).

Anstee, Joseph, 'Hanging By a Thread? China, America and the New Silk Road', *E-International Relations Students*, 2 March 2013. http://www.e-ir.info/2013/03/02/hanging-by-a-thread-china-america-and-the-new-silk-road

'Are We Entering a Golden Age of Gas?', Interview with IEA Head Maria van der Hoeven, *Economy Watch*, 13 February 2014.

Baev, Pavel, 'Russia Refocuses Its Policies in the Southern Caucasus', *Caspian Studies Program*, 2001 (International Peace Studies, Oslo).

Bahgat, Gawdat, 'Prospects for a Gas OPEC', *Middle East Economic Survey*, Vol. 52, Number 2, 12 January 2009, pp. 26–30.

Balmaceda, Margarita M., 'Ukraine's Energy Policy and US Strategic Interests in Eurasia', Woodrow Wilson International Centre for Scholars, Kennan Institute, *Occasional Paper 291*, May 2004. http://www.wilsoncenter.org/sites/default/files/op291_ukraines_energy_policy_balmaceda_2004.pdf

Barnett, Thomas P. M., 'India's Twelve Steps to a World-Class Navy', *Proceedings* (Annapolis, MD), July 2001.

Baumann, Florian, 'Energy Security as Multidimensional Concept', *Center for Applied Policy Research Policy Analysis*, Number 1, March 2008. http://edoc.vifapol.de/opus/volltexte/2009/784/pdf/CAP_Policy_Analysis_2008_01.pdf

Biresselioglu, Mehmet Efe, Muhittin Hakan Demir and Cansu Kandemir, 'Modelling Turkey's Future LNG Supply Strategy', *Energy Policy*, Vol. 46, 2012, pp. 144–152.

Blackwill, Robert D., and Meghan L. O'Sullivan, 'America's Energy Edge: The Geopolitical Consequences of the Shale Revolution', *Foreign Affairs*, Vol. 93, Number 2, March–April 2014, pp. 102–114.

Brenner, Michael, 'Geopolitics of Energy', 5 July 2012. http://energy.utexas.edu/the-geo-politics-of-energy/

Brower, Derek and Kirk Sowell, 'Pipelines, Not LNG, for East Med Gas', *Petroleum Economist*, Vol. 81, Number 6, July–August 2014, pp. 26–27.

Cain, Michael J. G., Rovshan Ibrahimov and Fevzi Bilgin, 'Linking the Caspian to Europe: Repercussions of the Trans-Anatolian Pipeline' (Rethink Institute, Washington, DC, *Paper 6*, 2012).

Caploe, David, 'Global Natural Gas Explosion: Clean, Cheap Product/Costly, Destructive Process', *Economy Watch*, 5 April 2011.

'Caspian Sea Oil and Natural Gas Exploration: Update and Hearing Summary', 27 June 1999 (American Geological Institute, Alexandria, 1999). www.agiweb.org/gap/legis106/caspian.html

Ceragioli, Paula and Maurizio Martellini, 'The Geopolitics of Pipelines', *Asia Times*, 29 May 2003.

Chan, Stephanie, 'Ain't No Mountain High Enough to Stop Pipeline Construction', *Pipeline International*, March 2012. http://pipelinesinterna tional.com/news/aint_no_mountain_high_enough_to_stop_pipeline_ construction1/067035

Chandra, Nayan, 'When Asia Was One', *GlobalAsia* (East Asia Foundation, Seoul), September 2006, pp. 58–68.

Chang, Felix K., 'Friends in Need: Geopolitics of China-Russia Energy Relations', *Orbis*, May 2014. http://www.fpri.org/articles/2014/05/ friends-need-geopolitics-china-russia-energy-relations

Chang, Gordon, 'China and Russia Form an Enduring Partnership', *China US Focus*, 28 May 2014. http://www.chinausfocus.com/foreign-policy/ china-and-russia-form-an-enduring-partnership/

Chaturvedi, B. K., 'Domestic Resourcing of Energy, Gulf and the Future of Global Energy', in Dr. S. N. Malakar, ed., *India's Energy Security and the Gulf* (Academic Excellence, Delhi, 2006), pp. 1–25.

Chen, Rosalie, 'China Perceives America: Perspectives of International Relations Experts', *Journal of Contemporary China*, Vol. 12, Number 35, May 2003, p. 287.

Cheney, Catherine, 'End of Egyptian Gas Deal No Threat to Israel's Energy Security', *World Politics Review Trend Lines*, 24 April 2012.

'China Begins Construction on 4,200 Kilometer Long Gas Pipeline Project', FBIS-CHI-2002-0704, 4 July 2002, Beijing Xinhua.

Chossudovsky, Michel, 'The Eurasian Corridor: Pipeline Geopolitics and the New Cold War', *Global Research* (Montreal), 22 August 2008. http://www.globalresearch.ca/the-eurasian-corridor-pipeline-geopolitics-and-the-new-cold-war/9907

Christoffersen, Gaye, 'China's Intentions for Russian and Central Asian Oil and Gas', *National Bureau of Asian Research Analysis*, Vol. 9, Number 2, 1998, pp. 1–34.

Clover, Charles, 'Dreams of the Eurasian Heartland: The Re-Emergence of Geopolitics', *Foreign Affairs*, March–April 1999. https://www.for eignaffairs.com/articles/asia/1999–03–01/dreams-eurasian-heartland-reemergence-geopolitics

Cordner, Lee, *Offshore Oil and Gas Safety and Security in the Asia Pacific: The Need for Regional Approaches to Managing Risks* (S. Rajaratnam School of International Studies, Monograph number 26, 2013).

Couloumbis, Theodor, 'Toward a Global Geopolitical and Geo-economic Concert of Powers?', *Southern European and Black Sea Studies* (London), Vol. 3, Number 3, 2003, pp. 17–27.

Crabtree, James, 'Asia Gas Buyers Club Threatens Canada's Energy Plans', *Financial Times Blogs*, 15 January 2014. http://blogs.ft.com/beyond-brics/2014/01/ 15/asia-wide-gas-buyers-club-throws- spanner-in-canadas-energy-plans

Dietl, Gulshan, 'Geopolitical Transformation in Central Asia: Implications for India', *Man and Development*, September 1995, pp. 145–169.

Dietl, Gulshan, 'The Security of Supply Issue: The Growing Dependence on the Middle East', in Pierre Audinet, P. R. Shukla and Frederic Grare, eds., *India's Energy: Essays on Sustainable Development* (Manohar Publishers and Distributors, New Delhi, 2000), pp. 209–223.

Dietl, Gulshan, 'Trends in Geopolitics of Energy: Stability in the Gulf', in Jasjit Singh, ed., *Asia's New Dawn: The Challenges to Peace and Security* (Institute for Defence Studies and Analyses, New Delhi, 2000), pp. 153–164.

Dietl, Gulshan, 'Remapping Global Energy: The Iranian Options', *GSP Occasional Paper Series* (New Delhi, January 2002).

Dietl, Gulshan, 'Stability in the Gulf: Implications for Energy Security', *Pacific and Asian Journal of Energy* (New Delhi), Vol. 12, Number 1, June 2002.

Dietl, Gulshan, 'Middle Kingdom and the Middle East: The Energy Connection', *China Brief*, Jamestown Foundation, Vol. 3, Number 4, 25 February 2003, pp. 61–69. www.jamestown.org/publications details.php?volume id=197issue id=669&article id

Dietl, Gulshan, 'New Threats to Oil and Gas in West Asia: Issues in India's Energy Security', *Strategic Analysis*, July–September 2004, pp. 373–389.

Dietl, Gulshan, 'Gas Pipelines: Politics and Possibilities', in I. P. Khosla, ed., *Energy and Diplomacy* (Konark Publishers, Delhi, 2005), pp. 74–89.

Dietl, Gulshan, 'Legal Regimes for the Caspian: The Iranian Position and Policies', in Kaw Mushtaq A. and Aijaz A. Banday, eds., *Central Asia: Introspection* (Centre of Central Asian Studies, Srinagar, 2006), pp. 189–196.

Dietl, Gulshan, 'Oil and Gas in the Gulf: The New Challenges', in S. N. Malakar, ed., *India's Energy Security and the Gulf* (Academic Excellence, New Delhi, 2006), pp. 103–115.

Dietl, Gulshan, 'Transnational Gas Pipelines: Global Context, Indian Experience', in Anwar Alam, ed., *India and West Asia in the Era of Globalisation* (New Century Publications, New Delhi, 2008), pp. 33–42.

Dietl, Gulshan, 'Transnational Gas Pipelines in Eurasia: Scramble for Resources and Routes', in Aijaz A. Bandey, ed., *Silk Route and Eurasia: Peace and Cooperation* (University of Kashmir Press, Srinagar, 2011), pp. 241–244.

Dietl, Gulshan, 'Musandam: Creating a New Region across the Water', in Steffen Wippel, ed., *Regionalising Oman: Political, Economic and Social Dynamics* (Springer, Dordrecht, 2013), pp. 279–287.

Dillon, Dana and John T. Tkacik Jr., 'China's Quest for Asia', *Policy Review* (London), December 2005–January 2006.

Dohmen, Frank, Alexander Jung and Jan Puhl, 'Stepping on the Gas: New Drilling Technologies Shake up Global Market', *Spiegel Online International*, 3 March 2011. http://www.spiegel.de/international/business/stepping-on-the-gas-new-drilling-technologies-shake-up-global-market-a-748573-2.html

Dorian, James P., Utkur Tojiev Abbasovich, Mikhail S. Tonkopy, Obozov Alaibek Jumabekovich and Qiu Daxiong, 'Energy in Central Asia and Northwest China: Major Trends and Opportunities for Regional Cooperation', *Journal of Energy Policy* (London), Vol. 27, 1999, pp. 281–297.

Dreyer, Iana and Gerald Stang, 'The Shale Gas "Revolution": Challenges and Implications for the EU', *ISS Alert*, February 2013. http://www.iss.europa. eu/uploads/media/Brief_11.pdf

Dunn, David Hastings and Mark J. L. McClelland, 'Shale Gas and the Revival of American Power: Debunking Decline?', *International Affairs* (London), Vol. 89, Number 6, 2013, pp. 1411–1428.

Edward, Matthew, 'The New Great Game and the New Great Gamers: Disciples of Kipling and Mackinder', *Central Asia Survey*, Vol. 22, Number 1, March 2003, pp. 83–102.

Engdahl, William, 'Syria Attraction: Russia Moving into Eastern Mediterranean Oil Bonanza', *RT*, 13 January 2014. http://rt.com/op-edge/syria-russia-war-oil-528

Erickson, Andrew and Gabe Collins, 'Beijing's Energy Security Strategy: The Significance of a Chinese State-Owned Tanker Fleet', *Orbis* (Philadelphia), Fall 2007, pp. 665–684.

Erickson, Andrew and Lyle Goldstein, 'Gunboats for India's New "Grand Canals"?', *Naval War College Review* (Newport), Spring 2009, pp. 43–76.

Escobar, Pepe, 'Iran Takes over Pipelineistan', *Asia Times*, 10 September 2005.

Escobar, Pepe, 'Playing Chess in Eurasia', *Opednews*, 25 December 2012. http:opednews/articles/ Playing-Chess-in-Eurasia-by-Pepe-Escobar-1112 25–365.html

Evans, Michael, 'Power and Paradox: Asian Geopolitics and Sino-American Relations in the 21st Century', *Orbis*, Vol. 55, Number 1, 2011, pp. 85–113.

Fan, Gao, 'Will There Be a Shale Gas Revolution in China by 2020?', *Oxford Institute for Energy Studies Paper*, Oxford, 18 April 2012. http://www.oxfordenergy.org/2012/04/will-there-be-a-shale-gas-revolution-in-china-by-2020/

Fedorenko, Vladimir, 'New Silk Road Initiatives in Central Asia', *Rethink*, Paper 10, August 2013. http://www.rethinkinstitute.org/wp-content/ uploads/2013/11/Fedorenko-The-New-Silk-Road.pdf

Foot, Rosemary, 'China and the United States: Between Cold and Warm Peace', *Survival* (London), Vol. 51, Number 6, 2010, pp. 123–146.

Fredholm, Michael, 'Globalisation and Eurasia's Energy Sector', *The Journal of Central Asian Studies* (Srinagar), Vol. 20, Number 1, 2011, pp. 1–17.

Friedman, George, 'The Geopolitics of India: A Shifting Self-Contained World', *Stratfor*, December 2008-April 2012.

'From the Editor: Energy Geopolitics in the 21st Century', *Journal of Energy Security*, Vol. 19, April 2012, p. 1.

'The Future of Natural Gas', *A Multidisciplinary MIT Study*, 2011. http:// mitei.mit.edu/system/files/NaturalGas_ExecutiveSummary.pdf

Gatemann, Reiner, 'Danes United behind the 'Most Ambitious Energy Plan in the World', *European Energy Review* (Groningen), 29 August 2013.

Gaynor, Daniel, 'Cybersecurity: Next National Security Challenge', *San Francisco Chronicle*, 4 June 2012.

'Geopolitics of Natural Gas', *Baker Institute Study*, Number 29, March 2005. http://www.bakerinstitute.org/publications/study_29.pdf

Ghorban, Narsi, 'The Evaluation of Recent Gas Export Pipeline Proposals in the Middle East', *Iranian Journal of International Affairs*, Vol. 7, Number 2, Summer 1995, pp. 449–465.

Ghorban, Narsi, 'Oil and Gas Pipelines from the Caspian Basin', *Amu Darya* (Tehran), Vol. 4, Number 1, Spring 1999, pp. 27–39.

Ghorban, Narsi, 'Iran's Potential Role in the Development and Utilisation of Oil and Gas of the Caspian Region', *Iranian Journal of International Affairs*, Vol. 12, Number 2, Summer 2000, pp. 264–273.

Ghorban, Narsi and Mohammad Sarir, 'Oil and Gas: An Outlook for Future Cooperation among the Persian Gulf States', *Iranian Journal of International Affairs* (Tehran), Vol. 5, Numbers 3–4, Fall–Winter 1993–94, pp. 738–753.

Glover, Peter C., 'Gas Discovery Changes Israel's Energy Picture', *World Politics Review*, 8 June 2009. http://www.worldpoliticsreview.com/articles/3885/gas-discovery-changes-israels-energy-picture

Goble, Paul, 'North-south Energy Routes More Attractive than East-West Ones', *Gas and Oil*, 24 January 2009. http://www.gasandoil.com/news/2009/02/ntr90740

Goodrich, Lauren and Marc Lanthamann, 'A New Era for Russia's Energy Strategy?', STRATFOR. Reprinted in *Economy Watch*, 14 February 2013. http://www.economywatch.com/economy-business-and-finance-news/a-new-era-for-russias-energy-strategy.14–02.html

Gorban, Alexander, Alexei Mastepanov and Alexander Orlov, 'Global Energy: New Geopolitical Equations', *International Affairs* (Moscow) Vol. 58, Number 6, 2012, pp. 179–192.

'The Great Land Robbery: Gwadar', *The Herald* (Karachi), June 2008.

Gretton-Watson, Paul, 'Energy: Wasted at the Wellhead', *Bulletin of the Atomic Scientists*, Vol. 58, Number 5, September 2002, pp. 22–23.

Grygiel, Jakub J., 'The Power of Statelessness', *Policy Review* (Washington, DC), April–May 2009.

Haghshenass, Fariborz, 'Iran's Asymmetric Naval Warfare', *Washington Institute for Near East Policy*, Focus 87, September 2008. http://www.washingtoninstitute.org/policy-analysis/view/irans-asymmetric-naval-warfare

Haider, Syed Fazl-e, 'A Great Game Begins as China Takes Control over Gwadar Port', *The National* (Abu Dhabi), 7 October 2012.

Hawdon, David, 'Market Competitiveness: The Economics of the Natural Gas Market and Its Competitiveness', in Stevens Paul, ed., *International Gas: Prospects and Trends* (Macmillan, Houndmills, 1986), pp. 14–33.

Holmes, James R. and Toshi Yoshihara, 'China and the United States in the Indian Ocean: An Emerging Strategic Triangle?', *Naval War College Review*, Summer 2008, pp. 41–60.

Hulbert, Mathew, 'Why China Will Stop US Energy Independence', *Forbes* (New York), 23 August 2012, p. 78.

Huntington, Samuel P., 'The Lonely Superpower', *Foreign Affairs*, Vol. 78, Number 2, March–April 1999, pp. 35–49.

Jaffe, Amy and David Victor, 'Geopolitics of Gas' (The James A. Baker III Institute for Public Policy, Rice University, *Working Paper Series*, 2004).

Jensen, James T., 'The Future of Gas Transportation in the Middle East Region: LNG, GTL and Pipelines', in *The Gulf Oil and Gas Sector: Potential and Constraints* (Emirates Center for Strategic Studies and Research, Abu Dhabi, 2006), pp. 249–281.

Jha, Saurav, 'Flush with Gas, Israel Now Must Find Ways to Export It', *World Politics Review*, 11 July 2013.

Kaplan, Robert D., 'The Coming Anarchy', *Atlantic Monthly* (Washington, DC), February 1994.

Kaplan, Robert D., 'Actually, It's Mountains', *Foreign Policy*, July–August 2010, p. 105.

Karbuz, Sobet, 'The Underbelly of Eastern Mediterranean Gas', *Journal of Energy Security*, 13 August 2013. http://www.ensec.org/index.php?catid=137:issue-content&id=445:the-under-belly-of-eastern-mediterranean-gas&tmpl=component&page&option=com_content&Itemid=422

Kaser, Michael, 'Soviet Gas Supplies', in Stevens Paul, ed., *International Gas: Prospects and Trends* (Macmillan, Houndmills, 1986), pp. 71–88.

Kelvin, T. Erickson, Ann Miller, E. Keith Stanek, C. H. Wu and Shari Dunn-Norman, 'Pipelines as Communication Network Links', *Science.gov*, 14 March 2005. http://www.osti.gov/scitech/servlets/purl/839987

Kemp, Geoffrey, 'The East Moves West: India and China's Great Game in the Gulf', *National Interest* (Washington, DC), Summer 2006. http://nationalinterest.org/article/the-east-moves-west-878

Kiernan, Peter, 'Pipeline Politics: Iran Looks to Send Natural Gas East', *World Politics Review*, 28 May 2008.

Klare, Michael T., 'The Bush/Cheney Energy Strategy: Implications for US Foreign and Military Policy', *Journal of International Law and Politics* (New York), Vol. 36, 2004, pp. 395–493.

Klare, Michael T., 'Energy: The New Thirty Years' War', *Guardian* (London), 29 June 2011.

Klare, Michael T. and Tom Engelhardt, 'The Tripolar Chessboard', *Tom Dispatch*, 16 June 2006.

Kolas, Ashild, 'Burma in the Balance: The Geopolitics of Gas', *Strategic Analysis* (New Delhi), Vol. 31, Number 4, July 2007, pp. 625–643.

Korin, Anne, 'Why Energy Terrorism Is Nothing New and Hard to Stop', *World Politics Review Trend Lines*, 2 May 2013.

Krauthammer, Charles, 'Who Made the Pivot to Asia? Putin', *Washington Post*, reprinted in *The Hindu*, 24 May 2014.

Kucera, Joshua, 'US Intelligence: Russia Sabotaged Pipeline ahead of 2008 Georgian War', *EurasiaNet*, 10 December 2014. http://www.eurasianet.org/node/71291

Kupecz, Mickey, 'Pakistan's Baloch Insurgency: History, Conflict Drivers and Regional Implications', *International Affairs Review* (Washington, DC), Vol. 20, Number 3, Spring 2012, pp. 95–110.

Ladislaw, Sarah O., Maren Leed and Molly A. Walton, 'New Energy, New Geopolitics: Balancing Stability and Leverage' (*A Report of the CSIS Energy and National Security Program and the Harold Brown Chair in Defense Policy Studies*, Center for Strategic and International Studies, Washington, DC), April 2014.

Lee, Bernice, Felix Preston, Jaakko Kooroshy, Rob Bailey and Glada Lahn, 'Resource Futures', (*A Chatham House Report*, December 2012). http://www.chathamhouse.org/sites/default/files/public/Research/Energy,%20Environment%20and%20Development/1212r_resourcesfutures.pdf

Leonard, Peter, 'China: Kazakhstan Unveil Landmark Gas Pipeline', *Associated Press*, 12 December 2009.

Levi, Michael, 'Splitting Rock vs. Splitting Atoms: What Shale Gas Means for Nuclear Power', *Bulletin of Atomic Scientists*, Vol. 68, Number 4, July–August 2012, pp. 52–60.

Lewis, Steven W., 'Russia to China Gas Deal: A Game of Spigots?', *Forbes*, 22 May 2014. http://www.forbes.com/sites/thebakersinstitute/2014/05/22/russia-to-china-gas-deal-a-game-of-spigots /

Luft, Gal, 'Building an Asian Energy Buyers Club', *Wall Street Journal*, 2 September 2013. http://online.wsj.com/articles/building-an-asian-energy-buyers-club-1409677827

Luft, Gal, 'The Sino-Russia Gas Deal', *Journal of Energy Security*, 27 May 2014. http://ensec.org/index.php?option=com_content&view=article&id=533:israels-zero-gas-option-take-ii&catid=143:issue-content&Itemid=435

Lukin, V., 'Energy: A Market Oriented Approach', *International Affairs* (Moscow), Vol. 54, Number 1, 2008, p. 73.

Lvov, Peter, 'Qatar's Great Power Games', *Oriental Review*, 30 March 2013. http://orientalreview.org/2013/03/30/qatars-great-power-games/

Mackinder, Halford, 'The Geographical Pivot of History', *Geographical Journal* (London), April 1904, pp. 421–37.

Mahalingam, Sudha, 'India-Central Asia Energy Cooperation', in Santhanam K. and Ramakant Dwivedi, eds., *India and Central Asia: Advancing the Common Interest* (IDSA and Anamaya Publishers, New Delhi, 2004), pp. 111–143.

Mahalingam, Sudha, 'Fuel of the Century or a Pipedream', *On the Brink: Desperate Energy Pursuits in South Asia* (Panos South Asia, Kathmandu), 2006.

Malik, Mohan, 'Energy Flows and Maritime Rivalries in the Indian Ocean Region', *Asia Pacific Centre for Security Studies* (Honolulu), 2008.

Maugere, Leonardo, 'Geopolitics of Energy: Discussion Paper 2012–10' (*Belfer Center for Science and International Affairs*, John F. Kennedy School of Government, Cambridge). http://belfercenter.ksg.harvard.edu/files/Oil-%20The%20Next%20Revolution.pdf

McKibben, Bill, 'Actions Speak Louder Than Words', *Bulletin of the Atomic Scientists*, Vol. 68, Number 2, March–April 2012, pp. 1–8.

Medlock, Kenneth B. III, Amy Myers Jaffe and Peter Hartley, 'Shale Gas and US National Security', *Baker Institute* (Houston), 19 July 2011, p. 12. http://bakerinstitute.org/files/496/

Meidan, Mical, 'The Implications of China's Energy-Import Boom', *Survival*, Vol. 56, Number 3, June–July 2014, pp. 179–200.

Mesquita, Bruce Bueno de and George W. Downs, 'Development and Democracy', *Foreign Affairs*, Vol. 84, Number 5, September–October 2005, pp. 78–85.

Mironova, Irina, 'A Sneak Peek into Russia's Energy Strategy up to 2035', *European Energy Review*, 30 January 2014. http://www.europeanenergyreview.eu/site/pagina.php?id=4250

Mitrova, Tatiana and Lakov Pappe, 'Gazprom: From the Big Pipe to the Big Business', *The Power of Oil and Gas* (Carnegie Moscow Centre), Vol. 10, Numbers 2–3. http://www.carnegie.ru/en/pubs/procontra/74445.htm

Mokhatab, Saeid, William A. Poe and James G. Speight, *Handbook of Natural Gas Transmission and Processing* (Elsevier, Burlington, 2006), p. 19. http://www.academia.edu/3126636/HANDBOOK_OF_NATURAL_GAS_TRANSMISSION_AND_PROCESSING

Momtaz, Djamchid and Saeid Mirzaee Yengejeh, 'The Legal Regime of the Caspian Sea: Iranian Perspectives', *Iranian Journal of International Affairs* (Tehran), Vol. 13, Numbers 2–3, Summer–Fall 2001.

Morozov, Yuri, 'Arctic 2030: What Are the Consequences of Climate Change? The Russian Response', *Bulletin of Atomic Scientists*, Vol. 68, Number 4, July–August 2012, pp. 22–27.

Morse, Edward L., 'Welcome to the Revolution: Why Shale Is the Next Shale', *Foreign Affairs*, May–June 2014. http://www.foreignaffairs.com/articles/141202/edward-l-morse/welcome-to-the-revolution

Nakano, Jane and Edward C. Chow, 'Russia-China Natural Gas Agreement Crosses the Finish Line', *CSIS*, 28 May 2014. http://csis.org/publication/russia-china-natural-gas-agreement-crosses-finish-line

Nathan, Hippu Salk Kristle, Sanket Sudhir Kulkarni and Dilip R. Ahuja, 'Pipeline Politics- A Study of India's Proposed Cross Border Gas Projects', *Energy Policy* (Elsevier, Philadelphia), Vol. 62, 2013, pp. 145–156.

Noel, Pierre, 'Asia's Energy Supply and Maritime Security', *Survival*, Vol. 56 Number 3, June–July 2014, pp. 201–216.

Nye, Joseph S. Jr., 'The Future of American Power: Dominance and Decline in Perspective', *Foreign Affairs* (New York), Vol. 89, Number 6, 2010, pp. 10–12.

Ocelik, Petr K. and Jan Osicka, 'The Framing of Unconventional Natural Gas Resources in the Foreign Energy Policy Discourse of the Russian Federation', *Energy Policy*, Vol. 72, 2014, pp. 97–109.

Odell, Peter R., 'Institutional Constraints on the Development of the Western European Natural Gas Market', in Stevens Paul, ed., *International Gas: Prospects and Trends* (Macmillan, Houndmills, 1986), pp. 89–106.

Odgaard, Ole and Jorgen Delman, 'China's Energy Security and Its Challenges towards 2035', *Energy Policy* (Amsterdam), Vol. 71, 2014, pp. 107–117.

Orlov, Alexander, 'Global Energy: New Geopolitical Equations', *International Affairs* (Moscow), Vol. 58, Number 6, 2012, pp. 190.

Pachauri, R. K., 'On Track with Iran: Shift in India's West Asia Strategy', *Times of India* (New Delhi), 19 April 2001.

Parfitt, Tom, 'Terror Alert as Georgian Pipeline Opens', *The Guardian*, 28 May 2006.

Pravosudov, Sergey, 'Lack of Russian Gas Strangles China', *Gazprom Magazine*, Number 6. http://www.gazprom.com/press/reports/2013/chinasuffocates

Raman, B., 'Hambatola and Gwadar: An Update', Chennai Centre for Chinese Studies, June 12, 2012, HYPERLINK "http://www.c3sindia.org" www.c3sindia.org. (accessed on 24 September 2016)

Rasizade, Alec, 'The Caspian Energy Legend and the "Great Game" of Concomitant Pipelines', *Iranian Journal of International Affairs*, Vol. 14, Numbers 1–2, Spring–Summer 2002.

Rasizade, Alec, 'The Mythology of Munificent Caspian Bonanza and Its Concomitant Pipeline Geopolitics', *Central Asia Survey* (London), Vol. 21, Number 1, 2002, pp. 37–54.

Riley, Alan, 'The Shale Revolution's Shifting Geopolitics', *The Hindu*, 31 December 2013.

Ross, Christopher, 'EU Energy Policy in the Caspian Region and Central Asia' (*Unpublished Paper*, International Conference on 'Energy, Transportation, and Economic Links in Eurasia: Emerging Partnerships' New Delhi, 16–17 January 2012).

Ross, Michael, 'The Political Economy of the Resource Curse', *World Politics*, Vol. 51, pp. 297–322.

Rumley, Dennis, Timothy Doyle and Sanjay Chaturvedi, 'Securing the Indian Ocean? Competing Regional Security Constructions', *Journal of the Indian Ocean Region* (London), Vol. 8, Number 1, June 2012, pp. 1–20.

Saivetz, Carol R., 'Caspian Geopolitics: The View from Moscow', *Brown Journal of World Affairs* (Providence), Vol. 7, Number 2, Summer–Fall 2000, pp. 53–61.

Sangkuk, Lee, 'China's Three Warfares: Origins, Applications and Organisations', *The Journal of Strategic Studies* (Oxon), Vol. 37, Number 2, 2014, pp. 198–221.

Sarir, Mohammad, 'Utilisation of Oil and Gas in the Caspian Region', *Amu Darya* (Tehran), Vol. 2, Number 1, Spring–Summer 1997, pp. 1–16.

Savin, Vladislav V. and Cherng-Shin Ouyang, 'Analysis of Post-Soviet Central Asia's Oil and Gas Pipeline Issues', *Journal of Eurasian Affairs* (Moscow), Vol. 1, Number 1, 2013, pp. 24–38.

Schweller, Randal L. and Xiaoyu Pu, 'After Unipolarity: China's Vision of International Order in an Era of US Decline', *International Security* (MIT Press, Massachusetts), Vol. 36, Number 1, 2011, pp. 70–72.

Shaffer, Brenda, 'Natural Gas Supply Stability and Foreign Policy', *Energy Policy*, Vol. 56, 2013, pp. 114–125.

Shahri, Nima Nasrollahi, 'The Petroleum Legal Framework of Iran: History, Trends and the Way Forward', *China and Eurasia Forum Quarterly* (Washington, DC), Vol. 8, Number 1, Spring 2010, pp. 121–122.

Sheridan, Greg, 'East Meets West', *National Interest*, November–December 2006, pp. 92–96.

Shivananda, H., 'China's Pipelines in Myanmar', *IDSA Comments*, 10 January 2012.

Siddiqui, Toufiq., 'Viable and Environment-friendly Sources for Meeting South Asia's Growing Energy Needs', *Asia Pacific Issues* (Honolulu), Number 83, August 2007, pp. 1–8.

Singh, Zorawar Daulet, 'China Deterrence Cannot Come from Navy', *The Hindu*, 7 August 2013.

Socor, Vladimir, 'Iran-Armenia Gas Pipeline: Far More than Meets the Eyes', *Eurasia Daily Monitor*, Vol. 4, Number 56, 21 March 2007, *The James Town Foundation*. http://www.jamestown.org/single/?no_cache=1&tx_ttnews%5Btt_news%5D=32607#.VTudizqJiUk

Stauffer, Thomas R., 'Caspian Fantasy: The Economics of Political Pipelines', *Brown Journal of World Affairs*, Vol. 7, Number 2, Summer–Fall 2000, pp. 63–78.

Stern, Jonathan, 'Gas OPEC: A Distraction from Important Issues of Russian Gas Supply to Europe', *Oxford Energy Comment*, February 2007. http://www.oxfordenergy.org/wpcms/wp-content/uploads/2011/01/Feb2007-GasOPEC-JonathanStern.pdf

Stevens, Paul, 'Pipelines or Pipe Dreams? Lessons from the History of Arab Transit Pipelines', *Middle East Journal* (Washington, DC), Spring 2000, pp. 224–241.

Subramanian, Arvind, 'The Inevitable Super Power: Why China's Dominance Is a Sure Thing', *Foreign Affairs*, Vol. 90, Number 5, pp. 66–78.

Sugar, R. D., 'Northrop Says Security Focus Pays Off', *Boston Globe*, 11 June 2004.

Tarock, Adam, 'The Politics of the Pipeline: The Iran and Afghanistan Conflict', *Third World Quarterly*, Vol. 20, Number 4, 1999, pp. 801–820.

Tashmuhamedova, N., 'Energy Transportation Routes in the Caspian and Central Asia', in P Stobdan ed., *Building a Common Future: Indian and Uzbek Perspectives on Security and Economic Issues* (Knowledge World, New Delhi, 1999), pp. 113–121.

Trenin, Dmitri, 'From Greater Europe to Greater Asia: The Sino-Russian Entente', *Carnegie Paper, Carnegie Moscow Center*, 9 April 2015.

Tsakalidou, Ilektra, 'The Southern European Corridor', *EU Institute of Security Studies, Issue Alert*, Number 21, July 2013, pp. 1–2.

Tucker, Paul W., 'The Natural Gas Market: The Cyclical Process', in Stevens Paul, ed., *International Gas: Prospects and Trends* (Macmillan, Houndmills, 1986), pp. 5–13.

Tunjic, Filip, 'War and Geopolitics – Really Together Again?', *Strategic Digest*, January 2000, pp. 20–34.

Tverberg, Gail, 'Behind Syria's Crisis: How Oil & Gas Limits Contributed to the Civil Unrest', *Economy Watch*, 11 September 2013. http://www.economywatch.com/features/syria-crisis-oil-and-gas-unrest.11-09.html

Ullah, Zabih, 'A View from Kandhar', in *Afghan Voices* (Lowy Institute for International Policy, Sydney), 15 December 2010.

Vaitheeswaran, Vijay V., 'Oil: Think Again', *Foreign Policy*, November–December 2007, pp. 24–30.

Varigonda, Keshav Chandra, 'The Pipeline That Wasn't: Myanmar–Bangladesh–India Gas Pipeline', *Journal of Energy Security*, 19 April 2012.

Vaughn, B., *Bangladesh: Political and Strategic Developments and US Interests* (Congressional Research Service, Washington, DC, R41194), April 2010.

Verma, Nidhi and Jo Winterbottom, 'Asian LNG Buyers Come Together in Bid to Secure Lower Prices', *Financial Post*, 3 December 2013.

Verrastro, Frank and Sarah Ladislaw, 'Providing Energy Security in an Interdependent World', *Washington Quarterly*, Vol. 30, Number 4, 2007, pp. 95–104.

Volkov, Vladimir, 'Gazprom: A Risky Strategy', *The Power of Oil and Gas* (Carnegie Moscow Centre), Vol. 10, Numbers 2–3. http://www.carnegie.ru/en/pubs/procontra/74445.htm

Wagner, Christian, 'Welcome to Interdependence: Energy, Security, and Foreign Policy in India' (*German Institute for International and Security Affairs*, Berlin, Working Paper, September 2006).

Walker, Martin, 'CHIMEA: The Emerging Hub of the Global Economy', *A. T. Kearney Report*, Washington, DC, 2008.

Walsh, D., 'Strategic Balochistan Becomes a Target in War against Taliban', *The Guardian*, 11 December 2009.

Weitz, Richard, 'Strategic Posture Review: Russia', *World Politics Review*, 1 February 2009.

Whitney, Mike, 'Et Tu, Mullah? Did Iran Just Knife Putin in the Back?', *Information Clearing House*, 18 August 2014. http://www.informationclearinghouse.info/article39446.htm

Wirsing, Robert G., 'Baloch Nationalism and the Geopolitics of Energy Resources: The Changing Context of Separatism in Pakistan', *Strategic Studies Institute* (Army War College, Carlisle, Pennsylvania), 17 April 2008.

Woods, Dwayne, 'Bringing Geography Back In: Civilisations, Wealth and Poverty', *International Studies Review* (Malden), Vol. 5, Number 3, September 2003, pp. 343–354.

Wong, Sterling, 'What's behind China's Global Oil and Gas Buying Spree? And Who's Next?', *Minyanville* (New York), 22 March 2013. http://www.minyanville.com/articles/print.php?a=48857

Yoshihara, Toshi and James Holmes, 'Command of the Sea with Chinese Characteristics', *Orbis*, vol 9, number 4, Autumn 2005, pp. 677–694.

Zambelis, C. 'Separatists, Islamists and Islamabad Struggle for Control of Pakistani Balochistan', *Terrorism Monitor*, Vol. 7, Number 37, 2009, pp. 9–12.

Web sources

http://pipelinesinternational.com (Melbourne)

http://www.iea.org/media/freepublications/AsianGasHub_WEB.pdf (International Energy Agency, Paris)

www.arielcohen.com (Ariel Cohen Website)

www.bakerinstitute.org (Baker Institute, Rice University)

www.bp.com (Hosted by the British Petroleum, also brings out Statistical Review of World Energy)

www.brooking.in (Washington, DC)

www.brookings.edu (Brookings Institute, Washington, DC)

www.chinausfocus.com (China-US Focus, Hong Kong)

www.csis.org (Centre for Strategic and International Studies)

www.dhs.gov (Department of Homeland Security, Washington, DC)

www.economywatch.com (Web Resource)

www.eia.gov (Energy Information Administration, Washington, DC)

www.eni.it (Hosted by the Italian Oil Company ENI, also brings out World Oil and Gas Review)

www.ensec.org (Journal of Energy Security)

www.eprints.lse.ac (London School of Economics)

www.eurasianet.org (New York)

www.ey.com (Earnest and Young)

www.gailonline.com (New Delhi)

www.gasandoil.com (Regularly updated and restricted site)

www.globalresearch.ca (Montreal)

www.iags.org (Institute for the Analysis of Global Security, Maryland)

www.iangv.org (International Association of Natural Gas Vehicles)

www.natgas.info (Hosted by Vivek Chandra, whose book is cited above)

www.oxfordenergy.org (Oxford)

www.pipelinesinternational.com (A Commercial Website from Houston, Texas)

www.rfa.org (Radio Free Asia, Washington, D C)

www.rfe/rl.com (Radio Free Europe/Radio Liberty, Washington, D C)

www.sage-india.com (South Asian Gas Enterprise)

www.setimes.com (Southeast European Times)

www.southasiamonitor.org (New Delhi)

www.stratfor.com (Web Resource)

www.teriin.org (The Energy and Resources Institute, also hosts Asian Energy Institute, a network of 16 energy institutes of Asian countries)

www.washingtoninstitute.org (Washington, DC)

www.worldbulletin.net (Istanbul)

www.worldenergy.org (World Energy Council with member committees in over a hundred countries.)

www.worldwatch.org (Web Resource)

INDEX

3 20